YOUR GUIDE
TO THE SKY

YOUR GUIDE TO THE SKY

RICK SHAFFER

Lowell House

• Los Angeles •

Contemporary Books

• Chicago •

Library of Congress Cataloging-in-Publication Data

Shaffer, Richard.
Your guide to the sky/Richard Shaffer.
p. cm.
Includes bibliographical references and index.
ISBN 1-56565-047-6
1. Astronomy—Observer's manuals. 2. Astronomy—
Amateurs' manuals. I. Title.
QB63.S555 1993
520—dc20   93-33660
                                                        CIP

Lowell House
2029 Century Park East, Suite 3290
Los Angeles, California 90067

Publisher: Jack Artenstein
Vice-President/Editor-in-Chief: Janice Gallagher
Director of Publishing Services: Mary D. Aarons
Project Editor: Bud Sperry
Cover and text design: Frank Loose Design

Photo acknowledgments: Cover, ul: NASA; ur: The Planetarium
Armagh–N. Ireland, ©Anglo-Australian Telescope Board; ll: D.
Malin, The Planetarium Armagh–N. Ireland, ©Anglo-
Australian Telescope Board; lr: D. Malin, The Planetarium
Armagh–N. Ireland, ©Anglo-Australian Telescope Board

Manufactured in the United States of America

10 9 8 7 6 5 4 3 2 1

Contents

Preface

My intention in writing this book was to provide a meaty beginner's guide to the sky without making the beginner think that she or he is being led to slaughter. I've included a lot of humor (or, what I allege is humor), and, hopefully a bit of the human side of amateur astronomy.

This book is based around 24 star maps. Since a lot of *zap* went into the star maps, I expended less zap on the other illustrations. To offset that, I've painted more than a few "word pictures" and made many appeals to the reader's imagination when describing some aspect of amateur astronomy that other books cover with illustrations. Also, I've not attempted to make this a comprehensive text on astronomy. What astronomy you'll find in this book is here to support your observing the sky. If you want a more comprehensive book on astronomy, one which features many illustrations, I recommend ASTRONOMY: A *Programmed Text* by Dinah Moche, published by Wiley.

Now I'll tell you a bit more about what this book doesn't feature, and why. I've limited the appendices to material I believe you'll actually use. This allowed me to devote the maximum number of pages to telling you about the Sky—what's out there, and why it's the way we see it.

Other books almost always feature an appendix that tells you where planetariums and observatories are, and what astronomy clubs might be in your area. There's usually a list of manufacturers of telescopes, and maybe, a list of publications. The common thread running through all these appendices is that they're obsolete long before the last copy of the book is sold!

Rather than include this guaranteed-obsolete information, I'll point you toward *Astronomy* and *Sky & Telescope* magazines. These two journals publish supplements to their magazines each spring that are radically better than any appendix I could con my editor into publishing. Each supplement provides up-to-date information on clubs, planetariums, observatories, manufacturers, dealers in equipment, star parties, and other meetings—the WORKS.

I hope this book kindles the spark that ignites your interest in astronomy. (The Bureau of Astronomical Hackneyed-Phrases required me to use just those words.) If it does, you'll want to obtain other books on astronomy. Rather than looking in an obsolete appendix, read the reviews of new books in the magazines I've already mentioned. Your fellow members of your local astronomy club are also an excellent source of recommendations on books.

Altadena, California
December, 1993

Acknowledgments

Anyone who tries to write a book without help is definitely in need of help—therapy, even. I may still be in need of both, but I did have a lot of help in writing this book.

First, I'd like to thank Jack Artenstein and Janice Gallagher of Lowell House for their initial faith in me that led to their decision to publish this book. When this first-time author and a first-time publisher of a book of star maps encountered considerable difficulties, Bud Sperry, my editor, showed enormous faith in somehow keeping the project going. Peter Hoffman, who supervised the production of the book, also showed considerable resourcefulness when the going got tough.

The raw star maps were generated using a special version of Etlon Software's excellent program *MacStronomy*. I am indebted to Stan Nolte, *MacStronomy's* creator, for providing this pre-release version of the program that allowed me to print out very "clean" star maps.

I created most of the final maps using a Mac IIci graciously loaned to me by my Frisbee-playing friend Blair Paulsen. While I ground out the maps, I was fed amazing vegetarian food by Blair's fiancée, Grace Roberts. I am eternally indebted to them both, as well as being "hooked" on Grace's fried-tofu-which-tastes-like-the-best-chicken-sandwich-you've-ever-tasted.

I thought I could design a good book until I saw the work of a real designer, Frank Loose. Frank's way of handling the text, tables, and illustrations is radically more attractive, friendly, and useable than my original conception of the book. It was also a genuine pleasure to work with him. (Lest you think I would ever attempt to make light of Frank's name, I point out that the section on page 125, entitled "A 'Loose' Star?", was written long before Frank was chosen to design the book.)

Richard Berry, Robert Burnham, and Steve Edberg spent many long hours going over the manuscript. I very much appreciate their comments and criticisms. Regardless of their help with this book, I greatly appreciate their friendship over the years. I've learned a lot from each of them.

When I needed two key illustrations, John Diebl of Meade Instruments and Douglas Knight of the Questar Corporation came to my rescue. John graciously provided Figure 10, a drawing of two typical beginner's telescopes, while Doug provided the Moon Map. I value both their assistance and their friendship.

Finally, I dedicate this book to my parents, John and Mary Shaffer. They certainly must have been astonished to have sired a loudmouthed, precocious, wanna-be astronomer. However, they never did anything to discourage my interest in astronomy. (That included allowing me and four friends to drive our family's only car, an old VW bus, from Chicago to Maine in 1963 to see my first total solar eclipse. I was only 17, and Dad drove a borrowed '47 Oldsmobile for 9 days. Occasionally, it even started.) Every budding astronomer should have parents such as these...

FINALLY, finally, this book bears my name. If you like it, I certainly expect to take a large share of the credit for it. If you find it wanting, I certainly want to hear about it. I just-as-certainly take sole responsibility for its faults.

Clear (and steady) skies...

CHAPTER 1

Introduction

1.1
What It Takes to Learn the Sky

If you spend a few hours with this book once or twice a month for the next year, you can learn to find your way around the night sky. Even if you don't use anything but your two eyes, you can get a lot of pleasure and knowledge out of learning the stars.

With just your eyes, you can find the constellations of the Zodiac and the one that represents your "sign." You'll be able to point out mighty Orion the Hunter and his two dogs, Canis Major and Canis Minor. You'll be sure that the dipper you're seeing is the Big Dipper. You'll find out how to know which of the stars is really one of the planets in our Solar System. If you should find yourself away from city lights, you can amaze your friends by pointing out an artificial satellite as it wends its way across the sky after sunset.

Instead of mistaking it for a cloud, you'll know that softly glowing band of light stretching from horizon to horizon is really the Milky Way, our home galaxy. With a little more practice, you'll be able to find the Andromeda galaxy and know you're looking at an object as it appeared more than two million years ago.

If you add a modest pair of binoculars to your stargazing arsenal, you can detect the craters of the Moon, four of the moons of Jupiter, and many of the brighter clusters of stars, clouds of gas and dust (nebulas), and galaxies that populate the night sky.

With a small telescope you can see the cloud belts of Jupiter, the rings of Saturn, and the phases of Venus. You'll also be able to distinguish many of the multiple stars (and know just which really is multiple). Finally, you'll get a much better view of all those clusters, nebulas, and galaxies than you could with your binoculars and you'll be able to see many more that were beyond their reach.

1.2
The Standard Warnings

1.2.1
A Very Serious Warning

Looking at astronomical objects is not dangerous, except for viewing that most dominating of all the objects in the sky, the Sun. So, every time I discuss observing the Sun, you'll find a more detailed version of this warning. Both my conscience and my lawyer demand that I make the following warning. For your safety and my peace of mind, please heed these warnings:

DO NOT EVER LOOK DIRECTLY AT THE SUN, EITHER WITH YOUR UNAIDED EYE OR THROUGH A TELESCOPE OR BINOCULARS, UNTIL YOU ARE THOROUGHLY FAMILIAR WITH THE SAFE METHODS OF VIEWING THE SUN. PERMANENT EYE DAMAGE OR BLINDNESS COULD RESULT.

DO NOT EVER LEAVE A TELESCOPE OR BINOCULARS UNATTENDED SO THAT A CHILD COULD POINT IT AT THE SUN AND LOOK THROUGH IT. BLINDNESS COULD RESULT.

1.2.2
A Slightly Less Serious Warning

Astronomy can be addictive! Once hooked, you'll be perfectly willing to totally upset your life and the lives of those around you, urging them to stay up until all hours of the night watching the sky with you. You'll spend much too much time poring over old issues of *Astronomy* or *Sky & Telescope* magazines. You might even spend more than a few bucks on a good telescope!

So, if you're perfectly happy watching wrestling or sitcoms every night on TV, put this book away. But, if you're ready to try something new, something that could turn into a lifelong, rewarding avocation, then read on.

1.3

How This Book Is Arranged

1.3.1

The Monthly Maps

This book is based on twelve maps of the night sky as observed from mid-northern latitudes. The maps are exact for any point as far north of the Earth's equator as Memphis, Tennessee, or Albuquerque, New Mexico. However, the maps will work just fine if you live anywhere in the northern hemisphere (U.S., Canada, Europe, north Africa, or Asia).

There are twelve "ALL-Sky" maps—one for each month of the year. Accompanying each ALL-Sky map is a ZOOM map that shows a smaller area of the sky in greater detail. Each ZOOM map is presented in larger scale than its corresponding ALL-Sky map.

Using the maps is easy: First, use the ALL-Sky map to locate the various groupings of stars. Then, use the ZOOM map to zero in on that portion of the sky best seen on the date and at the time you're out viewing.

A descriptive text accompanies each ALL-Sky map. It takes you around the sky in a different way each month. These Paths allow you to really learn the sky.

The text describing each month's ZOOM map tells you where to find interesting objects for viewing through binoculars or a small telescope. The descriptions of many of these objects will answer such questions as How far away is it? How big is it? Why does it glow the way it does?

1.3.2

The Explanatory Text

Even though the monthly sky maps and their matching descriptions occupy pages 94 through 160, I'm discussing them first. That's because I hope they will attract you both to this book and to the idea of learning the sky.

However, no knowledge is ever gained without cost. In this instance, you must read the first 93 pages of this book before you begin using the star maps.

1.3.3

The Appendix

The Appendix offers a variety of supplementary information. Although you can use the information in a stand-alone mode, the text on pages 3 through 93 will provide quite a lot of information on their use.

I think you'll find the mini-almanac particularly helpful. This appendix tells you the phases of the Moon, the locations of the easily seen planets, and the details of solar and lunar eclipses for every month between 1993 and 2000.

1.3.4

About the Outline Format

I've arranged this book in outline format with numbered sections. (The numbering scheme is identical to that used by lawyers. So, you see, lawyers have made at least *one* positive contribution to society!) I've done this to make it easy to find a particular section.

1.3.5

About Units, Mathematics, and Other Scary Stuff

Although astronomers use the metric system, most readers in the United States are more familiar with English measurement, so we'll use inches, feet, yards, miles, and so on. As they're needed, I'll introduce other units of measure and explain them.

If you want to become a professional astronomer, a thorough knowledge of mathematics is essential. However, mathematics is *not* required just to get to know the sky; for that you only need a desire to learn. I've kept mathematics, especially formulas, to a minimum.

1.4

If You're in a BIG Hurry, What to Read First

If you can't wait to get out under the stars, here is a short cut. Finish this chapter, then read the next two, and only the section in Chapter 4 on the human eye. Then read Chapters 5 and 6. By then you'll be familiar enough with the maps to begin. Once you've spent a few evenings under the stars, I'm betting you'll want to go back and read the rest of the chapters.

CHAPTER 2

How the Sky—and the Maps—Are Arranged

This chapter begins with what we all know best, the Earth on which we live, and expands your thinking from there.

2.1
How the Earth Is Laid Out

2.1.1
The Spherical Earth—A World (Global) View

Unless you're a member of the Flat Earth Society, you know our Earth is very close to being a sphere. You'll probably remember from school that a sphere is an imaginary surface surrounding an imaginary point so that any point on the surface of the sphere is the same distance from the center as any other point on the sphere. A beach ball is only roughly spherical, while a bowling ball is a much better sphere. A ball bearing is even better.

2.1.2
Magic and the Earth's Axis

To avoid confusing Moscow, Russia, with Moscow, Idaho, we long ago adopted a unique way of identifying any point on Earth. Its basis is the axis on which our planet rotates.

To understand an axis, think about Magic Johnson spinning a basketball on his finger. His finger supports the ball, but, to keep it from falling off, he must keep his finger exactly under the center of the spinning ball. The ball's axis is that imaginary line running from Magic's finger right through the center of the ball and out through the top.

We've defined the North and South Poles as the places where the Earth's axis intersects the surface of the Earth. Those two points do not move from side to side, they only rotate. If you stood exactly on Earth's North Pole, besides being cold, you'd only rotate round and round, and everything, including the sky, would rotate around you! (Also, every direction would be south!)

2.1.3
Don't Let the Generic Sphere Tie You in Knots

Instead of my trying to paint a word picture of all this, take a look at Figure 1. I call this diagram the Generic Sphere, because we'll be using it to help you visualize how the Earth is laid out and also the sky.

Many people are intimidated by diagrams like this. If you are, too, find a world globe and use it with the diagram. Don't feel inadequate if you have trouble with diagrams. I don't, but I am terrible at tying knots! I even bought an excellent book called the *Klutz Book of Knots*. It helped, but I still can't tie knots very well. Don't worry about the diagram, though; just find a world globe and muddle through. (And, if I were you, I wouldn't go mountain climbing with me!)

2.1.4
Using the Generic Sphere to Lay Out the Earth

2.1.4.1   Finding the Earth's Axis

The first thing you'll notice is a line running up and down between the letters N and S. That's the axis of the Earth's rotation. If you're going through

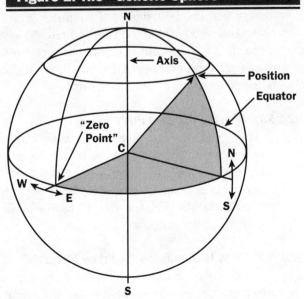

Figure 1: The "Generic Sphere"

this with a globe, that's the axis around which the globe spins.

2.1.4.2   Next, the Equator

The large oval with the letter C at its center is the plane of the Earth's equator, and the boundary line is the equator itself. Exactly half the Earth is north of the equator, the other half is south of it. Most world globes have a metal ring running along the equator. To be sure the line you've found is the equator, check to see that it runs through Kenya in eastern Africa and Ecuador in South America.

2.1.4.3   Then, the Meridians

The curves running from pole to pole represent the meridians. Every point on Earth has its own meridian, as the supply of these imaginary lines is, after all, unlimited. All points on the same side of the Earth that lie exactly north or south of one another lie on the same meridian.

2.1.4.4   Finally, the Parallels

The smaller oval that runs through the point marked "Position" represents a parallel. All points along a parallel lie due east or west of one another. The reason they're called parallels is that the plane formed by each parallel is parallel to the equator.

2.1.5
Using Meridians and Parallels to Define Position

Geographers uniquely define any position on Earth by using meridians and parallels. This system is connected to the positions of the stars and is very accurate. In fact, it's so accurate surveyors use it to ensure that the boundaries of countries, states, cities, and even your backyard are accurate. So, even if the river next to your backyard changes course over time, the mathematical position of your property doesn't. Let's see how it's done:

2.1.5.1   Finding the Zero Point

First, define the zero point. It's probably obvious from looking at Figure 1 that the zero point is the intersection of some meridian and the Earth's equator, but *which* meridian? The meridian of Earth's zero point runs through Greenwich Observatory in a suburb of London, England.

2.1.5.2   Why Is Greenwich the Prime Meridian?

Why? When the whole system was first being put into common use, the country whose sailors navi-

gated best was England, and the Royal Navy's observatory was located at Greenwich. So the rest of the world adopted Britain's methods of navigation and its Prime Meridian. (That England had the most powerful navy at the time might have had something to do with it.) If you're using a globe, find London and note that the Prime Meridian does, indeed, run right through the eastern suburbs of that city.

So the Prime Meridian runs through Greenwich and its intersection with the equator is Earth's zero point.

2.1.5.3   Longitude and Latitude

We define the position of each point on Earth by measuring how far its meridian is east of the Greenwich Meridian, and how far its parallel is north or south of the equator.

The measurement between the Prime Meridian and any other meridian is the *longitude* of that meridian.

The term used to describe the measurement between the equator and any other parallel is the *latitude* of that parallel.

2.1.5.4   Why Longitude and Latitude Are Measured as Angles

In our everyday lives, we give other people directions by using linear measure, like that made by a ruler or tape measure. We say: "The freeway on-ramp is two miles north on the left." Using miles or kilometers works perfectly well to give a person directions in our hometowns. But if you're on a ship out of sight of land, you need some other system. Geographers, navigators, surveyors, and astronomers use angular measurement to specify longitude and latitude. They do that because navigating across Earth's oceans would be impossible if they had to measure distance in miles or kilometers as they traveled across the sea.

Surveyors use their knowledge of the stars to make accurate determinations of position on land. Since they make their measurements of the positions of the stars as angles, it's convenient to think of longitude and latitude as angles.

2.1.5.5   About Angles

You probably already know what angles are, but it never hurts to be reminded. A circle is divided into 360 degrees, so a quarter of a circle equals 90 degrees, like the angle a carpenter's square makes. There's also an angle of 90 degrees between the horizon and the zenith or the "top of the sky."

2.1.5.6 Longitude as an Angle

Let's go back to the Generic Sphere again. The angle that is the longitude of point P is the lightly shaded, pie-shaped area in the plane of the equator. I've drawn it as an angle of around 90 degrees, but it could be any value from 0 to 360. That means a place only a bit WEST of Greenwich would have a longitude of more than 359 degrees, because we measure all longitudes EAST of the Prime Meridian.

2.1.5.7 Latitude as an Angle

The latitude of point P is the shaded slice of pie running from the plane of the equator north to point P. Latitude need not extend a full 360 degrees in order to fully define all points on Earth. Rather, it runs from −90 at the South Pole to +90 at the North Pole. So, any place with a positive latitude must be in the northern hemisphere and places with negative latitudes are found in the southern hemisphere.

2.1.6
Summary

The position of any point on Earth can be uniquely defined by knowing the angle of its meridian east of the Prime Meridian at Greenwich and by knowing the angle its parallel makes with the equator. Imagine these two angles, known as longitude and latitude, as you would see them from the center of the Earth.

2.2
Relating the Earth to the Celestial Sphere

We've spent quite a few words describing how the Earth is laid out. That's because we're about to take an enormous leap of faith or imagination to create the Celestial Sphere.

2.2.1
Step 1: A Journey to the Center of the Earth

This has nothing to do with B-movie versions of Jules Verne's sci-fi novel by that name. Rather, use your imagination to go to the center of our planet. It would be really helpful if you'd line up your body with the top of your head pointing north and your toes pointing south.

2.2.2
Step 2: See the Poles, Meridians, and Parallels

First, the line coming straight out the top of your head is the Earth's axis. All those lines running up and down on the Earth's surface are the meridians and the lines running horizontally around the Earth are the parallels. (If you want, imagine an extra meridian and an extra parallel running through your backyard. Remember, there is an infinite supply of these imaginary lines.)

2.2.3
Step 3: Explode the Surface of the Earth Outward

Here comes the leap of faith: Imagine that the stars and planets all lie on the surface of a sphere that's infinitely larger than the Earth. Call that new sphere the Celestial Sphere. Then, explode the Earth's longitude and latitude lines straight outward until they stick to the inside of the Celestial Sphere.

2.2.4
Step 4: Change the Names (to Protect the Innocent)

We've created this imaginary Celestial Sphere by exploding the longitude and latitude lines onto an infinitely large sphere. The Earth's equator becomes the celestial equator on the Celestial Sphere, and they share the same axis.

On the Celestial Sphere, Earth's meridians are called the lines of Right Ascension. So, on the Celestial Sphere, right ascension is the same as longitude on the Earth. The parallels of latitude on the Earth become parallels of Declination on the Celestial Sphere. So, just by specifying its right ascension and declination, we can specify the position of any point on the Celestial Sphere.

Just like longitude, right ascension is measured in degrees, minutes, and seconds east of its own "zero point" on the sky. The "zero point" is a point called vernal equinox. In Chapter 3, I'll explain why it's called that. In this system, one circuit around the Celestial Sphere equals exactly 24 hours of right ascension.

Just like latitude, declination is measured in degrees, minutes, and seconds north or south of the celestial equator.

2.3
How the Earth "Makes the Sky Move"

In the previous two sections, I introduced you to the system of determining the position of any place on Earth. Then we exploded the Earth, along with its axis, poles, equator, meridians, and parallels, into the infinitely large Celestial Sphere.

While you were thinking about how to create the Celestial Sphere, it was convenient to ignore the fact that the Earth is both rotating on its axis and revolving around the Sun. This, of course, makes the Celestial Sphere appear to rotate around the Earth.

2.3.1
About "Star-Time"

Suppose you point your telescope at a star just as it's crossing your meridian. Of course, if your telescope isn't moving relative to the Earth, the star will slowly move out of your field of view, because of the Earth's rotation. Now suppose you recenter the star in the field, and leave the telescope alone, maybe under the edge of your patio cover, until the Earth has completed one rotation. If you use a stopwatch to measure how long it took for the star to return to the field of your telescope, you'd find out that the interval was 23 hours, 56 minutes, and 3.56 seconds.

If your equipment is very accurate, you'll always get the above figure. Its highfalutin name is the *sidereal day*, but it's just as easy to refer to it as the "star day."

It's called the star day because that's the period it takes the Celestial Sphere to make exactly one circuit. (Of course, it's really the Earth that's rotated one circuit.) So, 24 hours of right ascension have slowly crawled through the eyepiece of your telescope in 23 hours, 56 minutes, and 3.56 seconds.

But doesn't the Earth revolve exactly once in a day? And isn't a day 24 hours long? Well, yes, but there are "days," and then there are "days." The "day" that elapses while the Earth turns once is the "star day" I mentioned, while we humans keep track of our daily lives using the "solar day." It's the solar day that's 24 hours long.

I'll tell you more in the next section about the solar day. Right now, we need to take care of a few more details.

The rotation of the Earth is what enables us to see a star at the same exact spot in the sky after 23 hours, 56 minutes, and 3.56 seconds, the star day. Since the stars are, effectively, an infinite distance away, the fact that the Earth also moves a tiny bit around its orbit in a day is of no consequence to the length of the star day.

However, the Earth's motion around the Sun does make the Sun appear to move around the Celestial Sphere once a year. Since the Earth moves from east to west around the Sun, the Sun appears to move from west to east around the Celestial Sphere. If the Sun weren't so bright, we could see it appears to move a tiny bit less than 1 degree eastward every day among the stars on the Celestial Sphere.

The effect we *can* see is that the stars rise about four minutes earlier each night than the night before. (They also set about four minutes earlier.) The precise difference between the rise time of a star tonight and its rise time tomorrow night is the difference in the length of the solar day and the star day: 24 hours minus 23 hours, 56 minutes, and 3.56 seconds. So, the difference is 3 minutes, 56.44 seconds, or about four minutes.

2.3.2
About the Solar Day

Before everybody spent much time worrying about anything except where to catch lunch, a practical concept of time was developed based on the motion of the most dominant object in the sky— the Sun. It rose every day, at noon it was highest in the sky, then it set at the end of the afternoon. In summer it rose higher in the sky and warmed us (sometimes too much), and in winter it didn't rise nearly so high as in summer, so we were cold. Although we knew the length of days changed with the seasons, we assumed that noon, when the Sun was highest, occurred at the same "time" each day.

Since the only method we had for measuring time was the Sun itself, using sundials or sun clocks, we never noticed that the Sun actually doesn't reach "high noon" at quite the same time each day.

Even before we had really accurate clocks, we had divided the day into 24 hours. But what was that "day"? Why, it was the time that elapsed between today's "high noon" and the next day's. Since our first crude mechanical clocks weren't very accurate, no one noticed the "high noon" problem.

When we began navigating long distances over the oceans, we developed more accurate nautical clocks called *chronometers*. During this era astronomers began to notice the more subtle aspects of time.

Chief among those details was that the length of the solar day varied gradually during a year. Two effects combine to vary the length of the solar day: One, the Earth's orbit is not perfectly circular, but elliptical or egg-shaped, with the Sun a bit to one side of the center of the ellipse. When the Earth is closest to the Sun it's moving a bit faster and when it's at its farthest, it's moving its slowest. That causes the solar day to be longer in January, when the Earth is closest to the Sun, than in July, when it is at its farthest.

Two, the Earth's axis is tilted with respect to the plane of its orbit. A whole chapter is devoted to this subject, so let's not cover it here.

The combination of these effects is not great, but it builds up day after day, then subsides, only to reverse itself. The result: "High noon" only occurs exactly at noon by the clock four days a year. At all other times, the Sun is either late or early, with the spread being as much as 14.5 minutes late and as much as 16.3 minutes early. Fortunately, this doesn't happen in a herky-jerky fashion, but varies smoothly. The pattern is essentially the same year after year.

Unless we're willing to adjust our clocks constantly to coincide with this variation in the solar day, we must invent a "fictional Sun" whose motion "averages out" the variations. Astronomers have done that. They call this the "Mean Sun," which moves exactly as the Sun would if Earth were in a perfect circular orbit and if its axis were not tilted with respect to the plane of its orbit.

Each day is a Mean Solar Day, and the Mean Sun arrives at "high noon" at precisely 12:00 every day. All this happens inside our clocks, regardless what the "real" Sun is doing. (In fact, just try to convince a clock of the existence of anything BUT the Mean Sun.)

2.3.3
About Time (Universal, Standard, and so on)

Since the Prime Meridian or zero point of longitude is at Greenwich, the Mean Solar Time at Greenwich was adopted as the standard of time throughout the world. And for a long time this standard time was called Greenwich Mean Time or GMT.

Recently, GMT was renamed UTC when astronomers refined its definition to include effects that are very small, but no longer insignificant in the Space Age. UTC is the French acronym for "Coordinated Universal Time." It's in French because the French operate the Bureau Internationale de l'Heure, the agency that coordinates all this time stuff, and incidentally, holds the hour (l'Heure) hostage. (The good folks at the BIH do a lot to make sure that the science of timekeeping is advanced in this increasingly sophisticated world.) Various government agencies broadcast accurate time signals (of UTC) over shortwave radio. In North America, you can hear the signals on frequencies of 5.000, 10.000, and 15.000 MHz.

Standard Time runs at the same rate as UTC. For everyone near the Prime Meridian, Standard Time and UTC are identical. So, noon in London is 12:00 UTC. However, the Sun can't be crossing the Greenwich Meridian at 12:00 UTC and also do high noon at Boston at 12:00 UTC. Instead, local noon in Boston occurs at roughly 17:00 UTC, about five hours later. But people in other parts of the world want noon (and 12 o'clock) to happen very near the middle of the day, regardless what UTC would indicate.

To accommodate traditionalists, who are certainly in the majority, there are 24 time zones, one of which is in synch with UTC, and the others as little as one hour and as many as twelve hours out of synch with the time at Greenwich. These zones are both numbered and named.

The number of each time zone represents the number of hours added to UTC to obtain the Standard Time for that zone. Since the Sun rises later for places west of Greenwich, Standard Time for any of those zones is earlier than UTC. A time zone west of Greenwich must be given a negative number so that when we add it to UTC, we come out with an earlier time. Likewise, if we know UTC and wish to know what time it is in Tokyo, east of Greenwich in the +10 time zone, we just add 10 to UTC to get Standard Time there.

Another way to think of this: For every 15 degrees of longitude east or west of Greenwich, the Standard Time changes plus or minus one hour. That's why the table of time zones for North America lists both the number of the zone and its central meridian:

Time Zones for North America		
Zone/Meridian	Place	Hours
Atlantic/60°W	Bermuda, Maritime Provinces of Canada, Puerto Rico	−4
Eastern/75°W	Montreal, New York, Atlanta, Miami, Panama	−5
Central/90°W	Winnipeg, Chicago, Dallas, New Orleans, Mexico City	−6
Mountain/105°W	Edmonton, Denver, Albuquerque, Baja	−7
Pacific/120°W	Juneau, Vancouver, San Francisco, Los Angeles	−8
Yukon/135°W	Canadian Yukon Territory	−9
Alaska/150°W	Alaska (except for Panhandle and Aleutians)	−10
Alaska/ Hawaii/165°W	Nome, Outer Aleutians, Hawaii	−11

During Daylight Saving Time, when computing UTC, add one hour to the zone number.

Why deal with all this time stuff, anyway? If you're only going to use this book to learn the constellations, you really don't need to know anything about time, except how to read your watch. However, if you're planning to observe an astronomical event that's predicted for a specific time, you'll want to be ready for it.

When you read in *Astronomy* or *Sky & Telescope* that an event will occur, the time mentioned for it will be UTC, not your Zone Standard Time. Since your non-astronomical life is governed by ZST, you live by that time. But if you need to convert the UTC of an event to your ZST, here's how:

Let's say you live in Truth or Consequences, New Mexico, and you know an eclipse of the Moon will begin at 03:17 UTC on January 15, 1996. Because New Mexico is in Mountain Standard Time Zone, the number you'll "add" to 03:17 will be −7. That would result in a negative time, so first you add 24 to 03:17, resulting in 27:17. Then you add 27:17 and −7 together, to get 20:17, or 8:17 P.M. Mountain Standard Time. Your zone time is always earlier than UTC, so the eclipse will occur at 8:17 P.M. MST on the 14th of January 1996.

O.K., so I cooked the books a little bit here. I set up the example so I could show you that events occurring early in the UTC day will often occur very late the previous day in your time zone if you live anywhere in North America. You have to be careful with time calculations if you want to get to an eclipse on time.

2.4
Showing the "Round" Sky on a Flat Page

2.4.1
About the Projection Used for Star Maps

In deciding the format for this book, I chose to provide you with simple, easy-to-use maps of the sky that would look as nearly like the actual sky as possible.

2.4.1.1 The Projection Shows Half the Celestial Sphere

This may seem obvious, but we can only see half the Celestial Sphere at any time, so I chose a projection that shows you only the half of the sky you can see. Choosing a projection that shows more

than half the sky would cause more distortion than we could tolerate.

2.4.1.2 A Projection That Minimizes Distortion

In fact, it's called the Airy Minimum-Distortion Projection, after Sir George Biddell Airy, the British astronomer and mapmaker who invented it.

2.4.1.3 Distortion Is Distributed Where It Will Do the Least Damage

Figure 2 illustrates what I mean. It's a map using the same projection I used for the monthly maps, except the stars have been omitted and meridians of right ascension and parallels of declination have been added.

The meridians are printed for each hour of right ascension, so there are 360 degrees per 24 hours, or 15 degrees between each hour. Likewise, a parallel of declination every 15 degrees has been printed, so very near the middle of the map the intersections of the meridians and parallels create a near square. If you look at the squares near the bottom of the map near the southern horizon, they're a bit distorted, but not so much you wouldn't call them squares. Also, the shapes of the squares near the North Celestial Pole (NCP) all look about the same.

Only when you look at the squares near the eastern and western horizons does their distortion become fairly extreme. Airy did that on purpose, because he wanted his map to minimize distortion near the middle of the map and near the southern horizon. The Airy Projection minimizes distortion along its central meridian.

Objects in the sky are best viewed when they cross your local meridian, because they're at their highest above the horizon and you're looking at them through the least atmosphere.

I chose the Airy Projection for these maps because it shows the constellation outlines with a minimum of distortion for the part of the sky from the southern horizon to the northern horizon along the map's central meridian. Since it distributes the distortion along both the eastern and western horizons, Airy's projection minimizes distortion compared to how great it would be if it were all distributed along one horizon.

I think you'll find that the shapes of the constellations in the maps will be remarkably close to those you'll see in the sky. Sometimes it will be wise to take into account distortion near the eastern or western horizon when you're learning a new constellation.

Figure 2: ALL-Sky Map Grid Lines

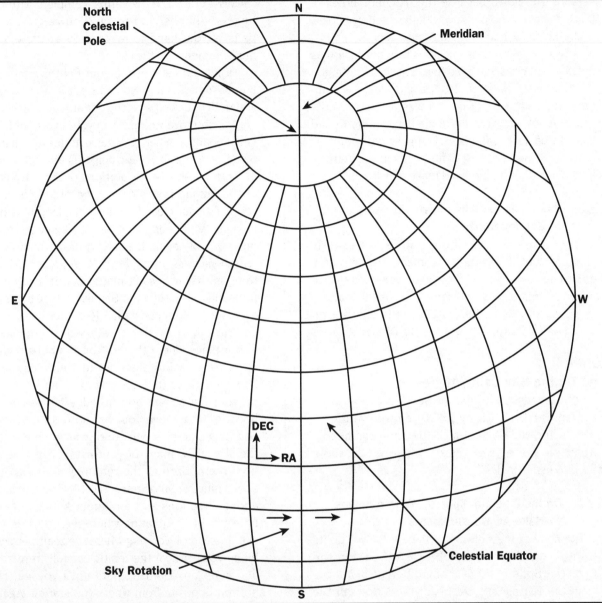

2.5

Using the Maps to Find the Stars

2.5.1
Which Way's North? or Getting Oriented

If you can't determine which way is north, the Sky Maps will be very difficult to use, so I've provided a list of ways for you to figure out which way is north. Here's the list:

2.5.1.1   Use a Map of Your City or Area

Most maps provide a compass rose, with arrows pointing to the cardinal points—north, east, south, and west. Find your street and compare it to the north-south line on the compass rose. In some places, streets run due north and south or due east and west. In others, they don't. Don't assume just because you live on North Main Street that Main Street runs due north and south. By carefully comparing the direction of your street to the north-south line, you can estimate the direction of true north.

2.5.1.2   Use a Compass

A magnetic compass consists of a magnetized pointer that has been mounted so it pivots freely horizontally above a scale. The scale is identical to the compass rose I just told you about, except it's usually divided into more directions. A scale of degrees is generally included, too.

The needle points to magnetic north, which is seldom true north because the magnetic poles of Earth do not line up perfectly with the rotational poles.

In the continental United States, central Canada, and Mexico, a compass indicates true north with as much accuracy as you need to orient the Sky Maps.

However, in western Canada and Alaska, a compass will point east of north by as much as 30 degrees, about a third of the angle between north and east. In eastern Canada, a compass will point as much as 30 degrees west of north. If you live in those areas, keep that in mind and make an allowance.

2.5.1.3   Ask Someone Who Knows the Stars to Get You Started

If you use this method, you're not exactly cheating, but if you do get someone to help you, don't have that person show you everything. If you do, you'll miss the fun of learning this stuff yourself. However, there's nothing wrong with being shown which way is north, or where the Big Dipper is.

2.5.2
Sizing Up the Sky Using Angles

We discussed angles in Section 2.1.5.5, and remember I mentioned a 30-degree angle in discussing the deviation of the compass needle from true north? Well, here's where we get more used to using angles.

2.5.2.1   On the Celestial Sphere, We Use Angles to Make All Measurements

Remember the Celestial Sphere is only a polite fiction. All the various objects we see on the Celestial Sphere are at many different distances from the Earth. But, except for the Moon, all the objects we see on the Celestial Sphere are so far away they might as well be infinitely far for all we can tell (without making extremely accurate measurements of star positions).

So we can't make any meaningful linear measurements of distances on the Celestial Sphere. We can't say "The bowl of the Big Dipper is 19 feet across." (We might as well say "The Big Dipper can hold 1600 gallons of water.")

But we can measure the direction of an object on the Celestial Sphere. To do that, we use angles.

2.5.2.2   Measuring Angles on the Sky

There are several ways to measure angles on the sky. Which one you use depends on how accurate you need to be.

2.5.2.2.1   Using Your Body Parts to Measure Angles

Most of us have two arms, complete with hands, fingers, thumbs, and so on. You can use various parts of your hand to make fairly accurate angular measurements on the sky.

Hold your fist close to your right eye so you can hardly see anything. Now move it slowly away from you until you can't stretch your arm any farther.

As you move it away from your eye, your fist gets "smaller and smaller." Of course, it doesn't change its actual size, the size that changes is its angular size.

Fortunately, the proportions of the various parts of the human body seem to be pretty constant. If you hold your closed fist at arm's length, it has an angular size of approximately 10 degrees. (The technical way of saying that is, "Your fist held at arm's length *subtends* an angle of 10 degrees.") The neat thing is that it doesn't matter if your people came from Angola, Lithuania, Shanghai, or Guatemala— everyone is equipped with a 10-degree fist.

You can prove this to yourself. Go outside and find a place where the horizon is unobstructed by large trees or buildings. Hold your closed fist at arm's length so its long side is up and down. Line up your fist so the bottom edge is even with the horizon. Now move your fist up from the horizon, one long side at a time, until you reach the "top of the sky." How many of your fists did it take to go from the horizon to the zenith? I'm betting it took about nine, or 90 degrees. (HINT: If you're as klutzy about directions as I am about knots, you might have trouble estimating where the zenith is. That's O.K. Instead of worrying where the zenith is, try this: When you get near the zenith, carefully turn around, and keep on counting fists until you get to the horizon opposite from where you started. Again, I'll bet you'll use about eighteen fists to get from one horizon to the opposite.)

You can make several approximate angular measurements using various parts of your hand. Just remember to hold it at arm's length:

Degrees	Body part (arm's length)
2°	Width of thumb
5°	Length of thumb
10°	Length of closed fist
15°	Length of closed fist, thumb extended

2.5.2.2.2   Using Your Binoculars to Measure Angles

You can use the field of your binoculars to measure angles on the sky. First, you need to know what the field is. The field in degrees is usually printed on the plate that also tells you the power, the size of the lenses, and the manufacturer.

Sometimes, the linear diameter of the field at a distance of 1000 yards is given. If that's the case, use the following table to determine the angular field of your binoculars:

Finding the Angular Field of Binoculars

Field (degrees)	Field (feet at 1000 yds.)
5°	262
7°	367*
10°	525
12°	631

* (Note: This is the field you'll find on the ZOOM maps.)

To measure distances on the sky, determine how many binocular fields separate the two objects in the sky you want to measure. To measure smaller angles, you'll have to estimate what fraction of the field of your binoculars separates the two objects, then multiply the fraction times the whole field.

2.5.2.2.3 Using Familiar Asterisms to Estimate Angles

An asterism is any group of stars that exhibit a distinctive pattern. The Big Dipper is the asterism most people know best and it's the one often used to demonstrate the approximate size of various angles on the Celestial Sphere. The angular distance across the bowl of the Big Dipper is 10 degrees and the bowl is 5 degrees tall.

2.5.3
Azimuth and Altitude

Even though the Celestial Sphere appears to rotate slowly, it's sometimes easiest to think about an object's position relative to your horizon, as if you were "freezing" the sky for a moment. You might think of this as the local sky. A position in the local sky can be defined by knowing its *azimuth* and its *altitude*

2.5.3.1 We Measure Azimuth Along the Horizon

The zero point for azimuth is due north. Azimuth is measured as an angle from north clockwise along the horizon. (The compass rose I mentioned earlier is an azimuth scale.) In the next column is a table of the various directions that correspond to various azimuths.

2.5.3.2 We Measure Altitude from Horizon to Zenith

Anything on the horizon has an altitude of 0 degrees, and anything at the zenith has an altitude

Azimuth Directions

Direction	Azimuth
North	0°
Northeast	45°
East	90°
Southeast	135°
South	180°
Southwest	225°
West	270°
Northwest	315°
North	360° (and back to 0°)

of 90 degrees. An object halfway up the sky from the horizon has an altitude of about 45 degrees.

2.5.3.3 Your Meridian Divides the Local Sky in Half

The local meridian extends from due north at the horizon, through the North Celestial Pole, through the zenith, to due south at the horizon. It divides the local sky in half.

2.5.4
Relating the Local Sky to the Celestial Sphere

We've spent a lot of time explaining how the motions of the Earth and Sun make the Celestial Sphere appear to rotate relative to the Earth. Now let's put all this together and get a feel for the results.

2.5.4.1 Journey to the Center of the Earth—Again

When you view the Celestial Sphere, it behaves exactly as if you were standing at the center of the Earth. All the stars appear to be moving in circles around the North Celestial Pole. The farther a star is from the pole, the larger and flatter the circle is.

Even though the sky rotates as if you were at the center of the Earth, you're obviously not there, and the Earth is obviously blocking your view of half the Celestial Sphere all the time. The horizon is the dividing line between Earth and sky. Because we humans are so incredibly small compared to the Earth, it looks to us like a flat plane slicing the Celestial Sphere in half.

2.5.4.2 The Stars Rise in the East and Set in the West

If you already either knew that or had figured it out from earlier sections, don't be offended. Just remember what a klutz I am when it comes to knots. I assume every reader needs to be told some of this stuff, so I'm including all of it.

If you look at Figure 2, you'll notice that the celestial equator and the parallels of declination look approximately like circles, but not quite. This is just another example of the distortion caused by projecting half the Celestial Sphere onto a flat page.

Those parallels show the paths the stars take in their daily journey through the sky. Stars that lie near the pole spend much more time above the horizon than those farther away from the pole.

2.5.4.3 Stars on the Celestial Equator Are Visible Half the Day

Those stars with a 0-degree declination lie exactly on the celestial equator. If you look at Figure 2, you'll see that the celestial equator meets the horizon at the due east and due west points. That's true no matter where you are on Earth. Stars on the celestial equator always rise due east and set due west. If you count the number of hours between the rise point and set point of the celestial equator, you'll notice there are twelve, or half of one day.

2.5.4.4 Stars South of the Celestial Equator Are Visible Less Than Half the Day

The farther south a star is on the Celestial Sphere, the longer it will be below the horizon each day. If you want to view certain objects that are very far south in the sky, you'll have to look at just the right time, or you'll miss them.

2.5.4.5 Stars North of the Celestial Equator Are Visible More Than Half the Day

Another look at Figure 2 will show you that the farther north a star is, the longer it will be above the horizon. These parts of the sky are easier to observe because they're above the horizon much longer than their southern cousins.

2.5.4.6 Stars Near the North Celestial Pole Never Set

Yet another look at Figure 2 shows you that near the North Celestial Pole there are complete parallels of declination that never meet the horizon. Stars inside these parallels are known as *circumpolar* stars. They never set and if the Sun didn't blot them out they would be visible all the time.

2.5.4.7 Polaris Is So Close to the Pole It Doesn't Appear to Move

Polaris is less than 1 degree from the pole, so it describes a tight little circle about the pole. Since it's so close to being fixed in the sky, it has long been used to point to north by explorers, navigators, and hikers.

2.5.4.8 The Portion of the Sky You See Depends on Where You Are

Get back into the imaginative mode you used when you first went to the center of the Earth, viewed the longitude and latitude lines from there, and then exploded them out onto the Celestial Sphere.

All right, you're at the center of the Earth, lying on a comfortable flat surface with your arms and legs spread out. The plane of your horizon is your blanket. Let's say an imaginary line runs at right angles to your stomach, right through your belly button. It always points to the zenith, at least while you're lying flat.

Since this is all imaginary, you can move your blanket any way you wish. First, orient it so it's parallel to the Earth's axis, with your head pointing north.

Now, "turn off the Earth," so all you can see is the Celestial Sphere. When you look out at the Celestial Sphere, you should see that your belly button/zenith line meets the Sphere at the celestial equator. As you watch, the stars are moving from your left to your right (from east to west).

To see how your horizon cuts the Celestial Sphere, extend your blanket out infinitely in all directions. Look due east (to your left) to see that the celestial equator runs straight up from the horizon and meets the zenith before plunging to the horizon at the due west point.

If you look back over your shoulder, you can see that the North Celestial Pole is right on the horizon at the due north point. The South Celestial Pole is at the due south point. Just for fun, trace your local meridian from the due north point through the zenith to the due south point. If you lie there for twelve hours, you'll be able to see the entire Celestial Sphere, although the Sun would blot out half of your view.

This imaginary trip is no different from what you could see from any point along Earth's equator. Only from the equator can you see the entire Celestial Sphere, although you would have to wait much longer than twelve hours to see all of the sky in darkness.

If you live north of the equator, you can never see a portion of the southern sky. Similarly, people who live in the southern hemisphere cannot see some of the northern sky, because the Earth gets in their way.

2.5.4.9 Lowest Possible Declination = Latitude –90 Degrees

If you live in Albuquerque at 35 degrees north latitude, you would add 35 degrees to –90 degrees to get −55 degrees. That the number is negative indicates the lowest possible declination you can see from Albuquerque is 55 degrees south of the celestial equator. Everything farther south is forever below your southern horizon.

2.5.5
The Monthly ALL-Sky Star Maps

Choosing how the monthly maps would be laid out was tough. We knew we wanted to cover the entire sky that could be seen at mid-evening on the fifteenth of each month. And we wanted a continuous map. Finally, we had to decide for which north latitude the maps would be exact.

2.5.5.1 The ALL-Sky Maps Show the Sky You'll See at 10:20 P.M. on the 15th of Each Month

We chose that time because that's when the even-numbered meridians of right ascension are just lined up with your local meridian.

There's a catch, though. The time by which you set your watch is your zone time, but the maps provide your local mean solar time, which depends on your longitude. If you live west of your time zone's central meridian, you'll have to wait a bit for the maps to represent exactly your sky. If you live east of the central meridian, you'll have to look a bit earlier. However, since the sky doesn't change very fast, you shouldn't have any trouble locating things.

You can also use each month's map at many other times throughout the year. If you buy this book in February and you want to start with January's stars, look for them at 8:20 P.M. on February 15.

Also, if you're out star watching on the first of the month, you can look for that month's stars at 11:20 P.M., or at the previous month's stars at 9:20 P.M. As you may have noticed from peeking at the maps, there's a lot of overlap from month to month. Think of them as "Cosmic reruns," and you'll feel right at home.

2.5.5.2 The ALL-Sky Maps Are Exact for Any Location at 35° North Latitude

If you live far north of latitude 35 degrees, you won't see the constellations at the very southernmost (bottom) part of the ALL-Sky maps. But you will see a bit more of the sky below the North Celestial Pole.

If you live far south of latitude 35 degrees, you can see some constellations not found on the maps. We had to compromise so we could ensure that the

maps were of greatest use to the greatest number of readers. (¡Que será, será! ¡Y, si usted vive al sur, usted necesita mapas de las estrellas en Español! ¿Verdad?)

2.5.5.3 Features of the ALL-Sky Maps

If you'll look at the ALL-Sky maps, I'll take you on a tour. The easiest way to see what's happening is to compare several of them. Let's start with January.

2.5.5.3.1 Circles, Cardinal Points, and Arcs

The circle that surrounds the map is your horizon. At four points around the horizon, the cardinal points (N, E, S, and W) are labeled.

The celestial equator is an arc that runs from due east, dips far toward the south, then rejoins the horizon at the due west point. To minimize clutter, the celestial equator is not marked on the ALL-Sky maps. It's identified on Figure 2, the "map grid."

2.5.5.3.2 Asterisms vs. Constellations

An asterism is merely a pattern of stars that's distinctive enough to be easily remembered. The ancients saw all manner of mythological characters and creatures in the asterisms they observed, and they gave the patterns names. Complex stories were associated with these named asterisms, many of them worthy of today's TV soap operas.

As learned men created and published maps of the stars, the named asterisms that became part of their maps achieved the status of constellations. Constellations have boundaries and all the stars within those boundaries are labeled with the name of that constellation.

Since the learned ones who made the maps were male, and therefore vain, they seldom resisted the urge to create their own constellations out of minor asterisms wedged in between the major ones. A common trick of a court astronomer on the outs with his monarch was to name a real (or imagined) asterism after the monarch. It might be said, then, that "a constellation is only an asterism with a good press agent."

As time passed, the sky became cluttered with constellations, and claim jumping was common. In 1925, to restore order, the International Astronomical Union adopted 88 officially recognized constellations and ended the squabbling.

The IAU assigned areas of the sky to specific constellations. There is no requirement that the constellation be easily seen or that the legends associated with them make any sense.

In fact, many of the constellations don't seem to resemble the creatures they represent, while others include easily seen asterisms as part of hard-to-see figures. Many books have been devoted to the figures of the constellations and how to see them.

2.5.5.3.3 About Constellation Figures

I consider knowing the constellations as merely one step in the process of learning the sky. I want you to find the stars and have a great time seeing them and everything else in the sky, so I've only drawn those constellation figures that make sense to me and to many other amateur astronomers I've been associated with over the years.

I've actually drawn a mixture of asterisms and constellation figures, so you'll find the outline of the Great Square of Pegasus instead of a fanciful winged horse. Unless you're an executive with a certain major oil company, I can't imagine you would see anything else in that part of the sky.

One weakness the ancients seemed to have was that when they ran out of ideas for naming constellations, they always seemed to fall back on the snake, serpent, or dragon, so there are no fewer than five such constellations, plus one "serpent handler." In keeping with the tradition of long, meandering constellations, we have Eridanus the River. As an example of the proposition that ANY nondescript group of stars can be a river, I drew the outline of Eridanus on the ALL-Sky for November.

2.5.5.3.4 Which Text Means What on the ALL-Sky Maps

I've used normal text of varying sizes to identify the constellations. Occasionally, I've identified individual stars that are parts of constellations nobody sees, but that need to be identified. These are in italics.

2.5.5.3.5 The Brightness of the Stars and Their Symbols

Near one corner of each ALL-Sky map is a block showing the symbols for the stars found on each map. Next to each symbol is a number ranging from −1 to 5. The general rule is, the larger the symbol, the brighter the star.

Those numbers refer to the magnitude of the star. The brighter it is, the lower its magnitude. That there are stars with negative magnitudes results from the history of the magnitude scale.

The same ancients we have to thank for the constellations grouped the stars into magnitudes. Since they had no instruments, they just eyeballed them!

When instruments came into common use for accurately measuring the brightness of stars, astronomers found that the stars commonly called "first

magnitude" covered a range of brightness much greater than those labeled second magnitude. In order to ensure the same mathematical relation between the brightness of each magnitude, some of the first-magnitude stars had to be labeled magnitude 0, and a few were promoted to −1.

That common multiplier between magnitudes is 2.512. Thus, a fourth-magnitude star is 2.512 times brighter than a fifth-magnitude star, and a third-magnitude star is 2.512 times brighter than a fourth-magnitude star, and so on.

The ALL-Sky maps in this book show stars to the fifth magnitude, even though humans can see to the sixth magnitude when the sky is very dark. I did this to avoid clutter. The ZOOM maps show stars to the sixth magnitude.

To find out how many times brighter the brightest star we can see is than the dimmest, we multiply 2.512 times itself eight times. (That's the same as "raising 2.512 to the 8th power.") The answer is about 1585.

You may have guessed that objects like the Sun and Moon and planets might have VERY negative magnitudes, and you'd be right. Here's a table of magnitudes:

Magnitudes	
Object	**Maximum Brightness**
Sun	−26.7
Moon	−12.6
Mercury	0
Venus	−4
Mars	−2
Jupiter	−2
Saturn	0
Uranus	6
Neptune	8
Pluto	14

2.5.6

Features of Monthly ZOOM Maps

Instead of cluttering the ALL-Sky maps with an incredible number of tiny symbols, there are twelve ZOOM maps, one for each month. These maps concentrate on a smaller area of the sky. The area that is the subject of each month's ZOOM map was chosen because it is in position to be viewed at its very best at the dates and times corresponding to the ALL-Sky map it accompanies.

2.5.6.1 Size of the ZOOM Maps Ranges from 50° to 70°

Each ZOOM map is big enough to include the best objects in its area, but no more. The legend on each map tells you the size of the map and shows you the relative size of the common 7-degree binocular field. Two maps have extra zoom fields to allow a clearer view of very crowded areas.

2.5.6.2 More Minor League Constellations Named

Since I am a worshiper of completeness, if not of obscure constellations, I've included the names of more minor league constellations on the ZOOMs than I have on the ALL-Sky maps.

2.5.6.3 More Stars Named or Numbered on ZOOMs

To help you learn the sky, all the first-magnitude stars found on these maps are named. Their names are in italics.

2.5.6.4 There Are Sometimes Additional Constellation Lines

For romantics, I've occasionally added additional constellation lines, not with any consistency, but then no one claims constellation lines are particularly consistent, either.

2.6
Your First Night Under the Stars

If you can't wait to get out under the stars, go ahead. But I hope you'll read the other chapters of this book sometime because they'll enhance your understanding of the sky and how to view it. This section is a catchall of material to get you started, although much of what you'll read here is covered in greater detail later.

2.6.1 Make an Astronomer's Flashlight

You want to be able to read the star maps, but if you use a light its glare will affect your view of the sky. What to do? Get a couple layers of red cellophane and tape them over an ordinary flashlight. Dim red light doesn't ruin your eyes' night vision, so you'll still be able to see the stars.

2.6.2 Pick a Moonless Night

You don't need any competition from the Moon, so unless you really feel the need for a challenge, don't go out for your first night of sky watching when the Moon is bright. (If it's only a little sliver, that won't dim your view of the sky.)

2.6.3 While Your Eyes Adapt to the Darkness, Just LOOK at the Sky

Even if you can't recognize a single constellation yet, spend some time while your eyes are "opening up" to enjoy looking at the sky. Depending on your world view, (1) it's good for your soul, or (2) it'll recharge your batteries.

2.6.4 If Possible, Recline on a Lawn Chair

A general rule is that, the more comfortable you are, the more you'll see. But don't get too comfortable or you'll fall asleep!

2.6.5 Don't Expect to See Everything the First Night

To have the best chance of finding a constellation or two your first time out, read the text accompanying the map you're going to use. There you will find a suggested "path" for finding the constellations of that month. Follow that path!

Even though the ALL-Sky maps don't change radically from month to month, the paths around the sky are usually quite different. If you follow a different path around the sky each month to learn many of the same constellations, you'll see them from a different perspective. My experience is that seeing patterns in the sky from several different perspectives helps you retain better what you've learned.

CHAPTER 3

The Path of the Sun, Moon, and Planets Through the Sky—the Ecliptic

The ecliptic has a chapter all its own because it's the Sun's path through the sky. The Moon and planets loiter along the ecliptic also, so it's a busy region of the sky.

3.1
Why the Ecliptic Is "Tilted"

In Section 3.4.3, I'll thoroughly discuss the constellations of the Zodiac. Here, I'll use them to illustrate that the ecliptic is tilted relative to the celestial equator. First, look at the ALL-Sky map for January. If you start at the eastern horizon, and work across the map to a point just south of the map's center, you'll notice that the constellations Leo, Cancer, and Gemini make a shallow arc up to that point. Then, Taurus, Aries, and Pisces continue the arc down to the western horizon. That arc is the ecliptic for the northern half of the sky. The celestial equator is south of the ecliptic as we see it on January's map.

Now look at the ALL-Sky map for July. Again starting at the east point on the horizon, note that an arc consisting of Aquarius, Capricornus, Sagittarius, Scorpius, Libra, and Virgo stretches across the sky toward the west point. However, this arc dips far south of its northern cousin. In fact, its maximum excursion is −23.5 degrees, in Sagittarius, and it reaches +23.5 degrees on the other side of the Celestial Sphere, in Gemini.

So, the plane of the ecliptic is tilted relative to the plane of the celestial equator. As you'll realize in a moment, what we see on the Celestial Sphere is, once again, not necessarily what's really going on…

3.1.1
The Ecliptic Isn't Tilted—the Earth's Axis Is

That shouldn't be too big a surprise, because I mentioned it in the previous chapter. However, it is of fundamental importance to our lives on Earth, as we shall see.

3.1.2
Earth's Axis Is Always Tilted 23.5 Degrees to the Plane of Its Orbit Around the Sun

Various theories attempt to explain why Earth's axis is tilted relative to its orbital plane. None of the theories manages to avoid having a serious flaw. What's important to us skywatchers is that the tilt of Earth's axis causes the Sun's path through the Celestial Sphere to be tilted 23.5 degrees to the celestial equator.

3.1.3
The Tilt of Earth's Axis Creates the Four Seasons

Look at Figure 3, which shows you the relationship of Earth's axis to the plane of its orbit. As this figure shows, the orientation of Earth's axis doesn't change as it circles the Sun. Only those four positions of the Sun with formal names are shown. To make the situations easier for you to visualize, I've drawn a flat Earth, although that may lead the FlatEarthers to try and quote me as an authority on the Flatitude of the Earth. Actually, the four Earths we see in Figure 3 represent a cross section of the Earth, what you'd see if you used your imagination to slice the Earth in half from pole to pole.

Even though we show the Earth at only four positions around its orbit, I'm sure you can imagine that it changes gradually from one situation to the next. These four situations become the four seasons.

3.1.3.1   Winter Solstice Is the Beginning of Winter

On December 22, the Sun fully illuminates the southern hemisphere, while part of the northern hemisphere is in darkness, even on the "daytime" side of the Earth. The Sun lies in the southern hemisphere of the Celestial Sphere and has reached the southernmost point in its travels for the year. If you went to a place at south latitude 23.5 degrees on the moment of the solstice at high noon, the Sun would lie at its zenith.

On December 22, that part of the northern hemisphere near the North Pole is in darkness all

Figure 3: Earth's Seasons

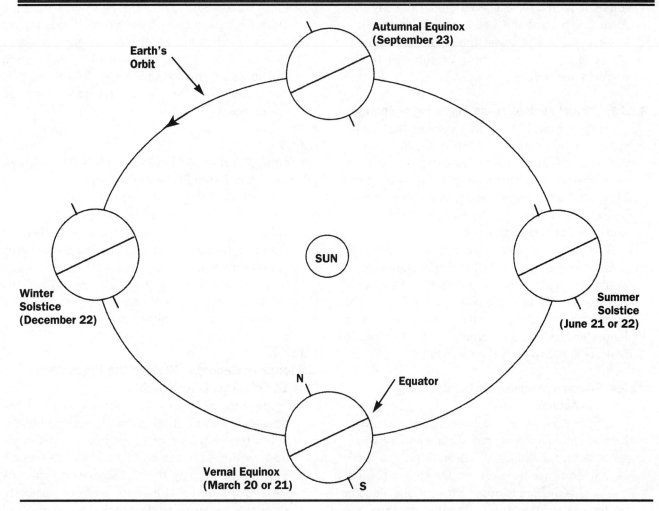

night and all day, while all of the area near the South Pole is in sunlight around the clock.

The areas where the Sun never rises for part of the winter or never sets for part of the summer are defined as the Arctic in the north and the Antarctic in the south. The dividing lines are called the Arctic Circle and the Antarctic Circle. You can find the lines on your globe at north and south latitudes 66.5 degrees.

Since the Sun is in the southern hemisphere of the Celestial Sphere, it is above the horizon in the northern hemisphere for the shortest time of the year on December 22. Therefore, it can't warm the northern half of the Earth as much as it can the southern. The Sun's rays are hitting the surface of the Earth at a steep angle in the northern hemisphere, but at a shallow angle in the southern. Therefore, less heat from the Sun hits each acre of the northern hemisphere than in the southern. These two effects combine to ensure that it's much warmer in the southern hemisphere in December than it is in the northern. I hope you've figured out

that on December 22, it's winter in the northern hemisphere and summer in the southern. By convention, we call the winter solstice the beginning of winter, even though that date is the one on which the northern hemisphere receives its minimum warmth from the Sun. That's because it takes the Earth's crust and atmosphere some time to radiate the heat of summer, so the meteorological seasons lag a bit behind the astronomical ones.

(Note: Most of the concepts covered here were developed in the northern hemisphere, so that perspective dominates the definitions. Although it's the beginning of summer in the southern hemisphere, everyone refers to December 22 as the winter solstice.)

3.1.3.2 Summer Solstice Is the Beginning of Summer

On June 22, we have the reverse of the winter solstice. Everything that described the situation in December is reversed, so the area above the Arctic

Circle is in daylight around the clock and the Antarctic is in darkness, even during the day. The Arctic has become the "Land of the Midnight Sun." The reason this is called the summer solstice is because it's the beginning of summer in the northern hemisphere.

3.1.3.3 Vernal Equinox Is the Beginning of Spring

Approximately halfway between the winter solstice and the summer solstice, the Sun crosses the celestial equator and passes into the northern hemisphere to begin spring. That happens on March 21, and is called the vernal equinox. At the precise moment of the equinox, the Sun's position is right ascension=zero, declination=zero.

If you were at Earth's equator at exactly high noon at the exact time of the vernal equinox, you would see the Sun exactly at its zenith. Likewise, if you stood at either of the poles, you would see the Sun exactly at the horizon. Everywhere else on Earth, the day would be evenly divided into twelve hours of light and twelve hours of darkness.

3.1.3.4 Autumnal Equinox Is the Beginning of Autumn

The autumnal equinox occurs approximately halfway between the summer solstice and the winter solstice on September 23. Everything about the vernal equinox is reversed for the autumnal equinox, as the Sun is crossing the celestial equator on its way south. At the autumnal equinox, the Sun's position on the Celestial Sphere is right ascension=12 hours, declination=zero.

3.1.4 The Angle of Tilt Doesn't Change Over Time, But the Direction of Earth's Axis Does Change (Slowly)

In Chapter 2 I said the Earth is not perfectly spherical. Rather, Earth is a bit pear-shaped. For the record, the Earth is approximately 26 miles wider at the equator than at the poles. Our home planet is also a bit wider south of the equator than north of it.

If Earth were perfectly spherical, its axis of rotation would always point to the same point on the Celestial Sphere. Since that's not the case, the Sun, which is massive, tweaks the direction of Earth's axis, so over time it sweeps out a cone. It does this just as the axis of a kid's top sweeps out a cone, only radically slower. In fact, Earth's axis takes about 26,000 years to complete one revolution around the cone. This effect is called *precession*, and we say the Earth's axis precesses as it rolls around the cone.

The Moon also tweaks the direction of the Earth's axis, but its effect is much less than the Sun's. This effect takes only 18.6 years to complete one cycle. When you add the Moon's effect (which is called *nutation*) to the precession of the axis caused by the Sun's gravity, the shape the Earth's axis sweeps out looks like a cross between a funnel and a jello mold.

3.1.5 In Terms of a Human Lifetime, Earth's Axis Always Points to the Same Place in the Sky

You'll probably remember that the direction of Earth's axis fundamentally defines the Celestial Sphere. If the axis were precessing quickly, then the Celestial Sphere would also be changing quickly. Fortunately, that's not the case. Rather, Earth's axis takes 26,000 years to precess just once. Thus, the Celestial Sphere we see tonight is the same as the one we see tomorrow night, and so on.

3.1.6 In Terms of Recorded History, the Precession of the Earth's Axis Is Noticeable

Because of precession, Earth's axis did not point near Polaris at the dawn of recorded history. When the Pharaohs of Egypt had the pyramids built, a dim star named Thuban was the North Star. Archeologists theorize that the pyramids were built using the stars as very accurate standards because the descending access shaft of the pyramid at Gizeh points exactly where Thuban would have passed when it was just below the North Celestial Pole.

That was 4800 years ago. In another 21,000 years, Thuban will once again be the North Star. If we and the pyramids survive that long into the future, things will once again be aligned as they were in those mysterious times.

3.1.7 Precession Is Also Noticeable in Terms of Measuring Accurate Positions on the Celestial Sphere

You may have noticed I spent considerable time discussing right ascension and declination, the two angles we use to measure position on the Celestial Sphere, but failed to mention where the zero point of right ascension is. Astronomers define zero hours of right ascension as the point where the celestial equator meets the ecliptic, at a place called the vernal equinox.

Since Earth's axis is precessing, it's taking the Celestial Sphere with it, and it's dragging the zero point of right ascension along for the ride. Think of the whole Celestial Sphere as slowly rotating along the ecliptic, with the same period of rotation—26,000 years. When you think of it that way, the North Celestial Pole is rotating around the pole of the ecliptic, taking the same 26,000 years to do it.

A reasonable question might be, "If the intersection of the ecliptic and the celestial equator is doing all that precessing and nutating, then why pick it as the zero point?" Because everything is moving. The advantage of the vernal equinox as the zero point is that its motion is reasonably predictable, unlike other candidate zero points.

3.2

Motion of the Sun Along the Ecliptic

Before we begin, here's your occasional dose of reality: Even though most of the time I'm writing about the motion of the Sun along the ecliptic, what's really happening is that Earth is orbiting the Sun. The Celestial Sphere is just a polite fiction, along with the mathematical lines that define the celestial equator and the ecliptic. (However, Earth's equator MUST be real. After all, Ecuador's named for it!)

3.2.1

The Sun Doesn't Move Uniformly Along the Ecliptic

All the planets move around the Sun in an orbit that's called an *ellipse*. An ellipse is a path that's oval. Some elliptical orbits are nearly circular, while others are long and narrow. The *eccentricity* of an ellipse is the measure of how close to circular an elliptical orbit is. If the eccentricity is zero, the orbit is circular. If the eccentricity is between 0 and 1, the orbit is an ellipse. If its eccentricity is 1, the orbit is infinitely large, and is called a *parabola*. Such an orbit is open on the far end.

An elliptical orbit is said to have two foci, two points located along the line that bisects the long axis of the ellipse. The massive body in a system occupies one of these foci. The more eccentric an elliptical orbit, the closer the foci will be to the ends of the ellipse.

When a planet in an elliptical orbit is at its closest point to the Sun, we say it's at *perihelion*. The opposite, the farthest point, is called the *aphelion*. A

planet has its highest velocity at perhelion, its lowest at aphelion.

Technically, the Sun and the planets orbit a common center. However, since the Sun is radically more massive than any other body in the Solar System, we usually think of the Sun as being fixed, with all other objects in the system in orbit around it.

The eccentricity of Earth's orbit is 0.07, which makes it nearly circular. Only Venus and Neptune are in orbits that are closer to circular than Earth's.

Earth moves a bit faster near perihelion, its closest point to the Sun, in January than in July when it is at its farthest point, the aphelion. So, it takes less time for the Sun to traverse the southern hemisphere in its travels than it does to traverse the northern hemisphere. That the northern winter is shorter than its southern counterpart suggests it would be milder than the southern winter.

Since Earth is three million miles closer to the Sun in January than in July, we receive more warmth from the Sun in the northern winter than we do during the southern winter. The inverse is true also. In the northern summer, we receive less warmth from the Sun than in the southern summer.

It would appear, then, that in the southern hemisphere winters would be harsher and the summers hotter than the same seasons in the northern hemisphere. If the two hemispheres were identical that would be the case, however, much more of the southern hemisphere is covered by water than the northern. That additional water seems to serve as a thermal flywheel, absorbing extra heat in the southern summer and releasing it in the winter. Therefore, the corresponding seasons in both hemispheres seem about the same.

3.3

Motion of the Moon Along the Ecliptic

Luna, or as it is better known, the Moon, is our closest neighbor in space. The plane of the Moon's orbit is inclined about 5 degrees with respect to the plane of Earth's orbit. When that's projected onto the Celestial Sphere, the Moon's path closely follows the ecliptic, never straying more than 5 degrees north or south. The Moon can travel as far north as +28.5 degrees declination or as far south as −28.5 degrees.

Just as the ecliptic extends half above the celestial equator and half below, the Moon's path extends half north of the ecliptic and half south. The points where the Moon's orbit crosses the ecliptic are called nodes. The ascending node is where the Moon crosses the ecliptic going north, and the descending node is where the Moon crosses the ecliptic going south. The line connecting the two nodes is called the line of nodes.

The Moon's orbit is tweaked by the Sun's gravity so that the line of nodes slowly precesses westward. It takes 18.6 years to complete one cycle of the precession of the nodes.

3.3.1
Moon's Orbit

The Moon orbits Earth at a distance of approximately 239,000 miles. Its orbit is a bit elliptical, too, so it can come as close as 226,000 miles and recede as far as 252,000 miles. The terms used to describe the closest and farthest points of any satellite from Earth are *perigee* and *apogee*.

The period of the Moon's orbit is 27.32 days. That is, it takes the Moon that long to return to the same position along the ecliptic. However, it is the *synodic period* of the Moon that is of greater interest, since that is the period that elapses between successive conjunctions with the Sun.

A *conjunction* occurs when two objects are lined up with our line of sight on Earth. When the Moon is lined up with the Sun, we say it is "in conjunction with the Sun." That's the technical term, but it's much more commonly known as "new moon." The period between new moons, or the Moon's synodic period, is 29.53 days, which is often called the *lunar month*. It's also called a *lunation*.

3.3.2
Phases of the Moon

During one lunar month, the Moon goes through all its phases. Figure 4 shows you why the phases of the Moon are as they are. Just like Earth, half the Moon's spherical surface is lit by the Sun, the other half is in darkness. The phases we see result from the angle the Moon makes with the Sun.

At new moon, the Moon is lined up with the Sun from our view on Earth. We see the dark side of the Moon or, rather, we don't see the Moon at all, because the extreme brightness of the Sun outshines the extremely dim Moon. What happens when the Moon is EXACTLY lined up with the Sun is described in Section 3.3.3.1.

As the Moon moves eastward away from the Sun in the sky, we see a bit more of its sunlit side each night. So, a few days after new moon, we can see a thin crescent in the western evening sky. (At this point, and for the next few nights, you can see the dark side of the Moon faintly illuminated by sunlight reflected off the Earth. This is called *Earthshine*. It's especially beautiful viewed through binoculars.)

The crescent Moon continues to *wax*, or grow fatter. When half the Moon's disc is illuminated, we say the Moon has reached first-quarter phase. That term comes from the fact that the Moon is a quarter of the way through lunation. At first quarter, the Moon is 90 degrees east of the Sun along the ecliptic, so we're looking at the sunlit side of the Moon from off to the side. Because it is 90 degrees from the Sun, the first-quarter moon sets approximately halfway through the night.

The Moon continues to wax, but it's no longer a crescent. During this period after first quarter, we say we have a *gibbous* Moon. Once it is opposite the Sun in the sky, we see all of the Moon's sunlit side. We call this—ready?—full moon. The full moon is 180 degrees around the ecliptic from the Sun in the sky, so it rises almost exactly as the Sun is setting, and sets almost as the Sun rises the next morning.

The second half of the Moon's journey is the inverse of the first. Now the Moon is waning, or growing slimmer, and it's once more described as gibbous. Each evening it rises later and later after sunset.

Three-quarters of the way around its journey, the Moon once again shows us one side of its disc illuminated and the other in darkness. However, the illuminated side we see now is the one that was in darkness at first-quarter phase. The Moon rises around midnight and crosses the local meridian around sunrise.

As it approaches new moon, the phase of the Moon is a waning crescent, until we can't see it at all. It becomes a new moon, then repeats the cycle.

Because its orbit moves the Moon eastward among the stars, the Moon rises later each night than the night before. If you timed the rising of the Moon each day for a month, you'd find it rises on the average fifty minutes later each night than the night before, but the variation can be as much as plus or minus fifteen minutes.

The average period between successive appearances of the Moon is about 24 hours, 50 minutes. If you're associated with the nautical world, you might notice that figure is just twice 12 hours, 25 minutes, the average time between successive high

Figure 4: The Phases of the Moon

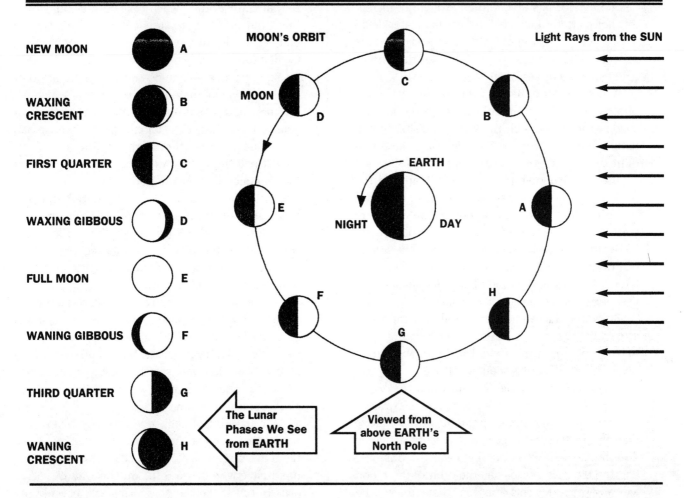

NEW MOON A

WAXING CRESCENT B

FIRST QUARTER C

WAXING GIBBOUS D

FULL MOON E

WANING GIBBOUS F

THIRD QUARTER G

WANING CRESCENT H

MOON's ORBIT

MOON

EARTH

NIGHT DAY

Light Rays from the SUN

The Lunar Phases We See from EARTH

Viewed from above EARTH's North Pole

tides. This should come as no surprise, because the Moon's gravity is the primary cause for the tides. (The Sun is the other serious player in the tide game.)

You can demonstrate the phases of the Moon for yourself by using a lamp and a baseball. Place the lamp with its shade removed in one end of an otherwise darkened room. Sit in the other end of the room and hold up the ball in front of you between your face and the lamp. Now revolve the ball around your head, holding it at arm's length. (A rotating desk chair helps.) Do this slowly and revolve from right to left, which is the same as from west to east on the Celestial Sphere. As the baseball orbits your head, you can see it go through the same phases as the Moon. You can even demonstrate the next phenomenon we'll cover—eclipses!

3.3.3
About Eclipses

If the plane of the Moon's orbit were not inclined to the plane of the Earth's orbit, the Moon would

move exactly along the ecliptic, and would pass between the Earth and Sun every new moon, and through Earth's shadow every full moon. These two types of events are known as *eclipses*. The ecliptic got its name because these events happen on or near it on the Celestial Sphere. Because the planes of the two orbits are slightly inclined, eclipses happen less often. That's a pity, because eclipses are some of the most interesting events in the sky.

3.3.3.1 How Solar Eclipses Happen

So far, we've only discussed angular size in degrees and decimal fractions of a degree. To discuss eclipses of the Sun, we need a finer division. So, we'll divide each degree into 60 minutes of arc, and divide each minute into 60 seconds of arc. An angular measurement of 17 minutes of arc is written 17', while 17 seconds of arc is written 17".

The Moon is about 31' in apparent diameter, while the Sun is a bit smaller in apparent diameter. Thus, when the Moon passes between Earth and Sun, all of the Sun's "surface" can be covered by the

Moon. (I put "surface" in quotes because the Sun has no surface. It's an enormous sphere of incredibly hot matter, which we'll discuss in greater detail in Chapter 7. The layer of matter we see is called the photosphere, and it's the diameter of the photosphere that's a bit less than 31'.)

While the Moon covers the extremely bright photosphere of the Sun, observers in the Moon's shadow can see possibly the most exquisite sight in nature, the Sun's corona. This sight and the attendant experience of a total solar eclipse attract thousands of people to those places around the world where the Moon's shadow touches the surface of the Earth. If ever you can, observe an eclipse of the Sun. Only be warned: For many people, watching eclipses is definitely addicting! Also, *be warned that viewing an eclipse with the unprotected eye when the Moon is NOT covering ALL the Sun is no less dangerous than viewing the Sun unsafely at other times.*

Since the Sun and Moon are almost the same apparent size, the conditions under which the eclipse occurs can affect the kind of eclipse observers in the Earth's shadow will see.

A *total eclipse of the Sun* occurs when the Moon completely covers the Sun's photosphere, exposing the corona. If the Sun's apparent size is exactly that of the Moon's, the period of totality lasts only a moment. If the distance between the observer and the Moon is as close as the orbit of the Moon will allow, the Moon's apparent diameter will be larger than its average value, and totality will last much longer. Not only does the position of the Moon in its elliptical orbit affect the duration of the eclipse, but the geometry of the eclipse also affects the duration. If observers are in the Moon's shadow near sunrise or sunset, they are almost 4000 miles farther from the Moon than someone standing in the shadow near high noon. The possible duration of totality varies from a split second to about 7.5 minutes.

If conditions conspire to place the Moon far enough from the observer standing in its shadow, the apparent size of the Moon will not be adequate to cover the Sun's photosphere. An *annular eclipse* is the result. It's called that because the observer sees an annular ring of the Sun around the dark Moon. This ring is sometimes called a "ring of fire." Because part of the photosphere is uncovered by the Moon, the corona remains invisible to observers of an annular eclipse.

If the Moon's shadow just misses the surface of the Earth or if the observer is not quite in the shadow, he or she sees a *partial eclipse of the Sun*. In a partial eclipse, the Moon "takes a bite out of the Sun," but never completely covers it. This type of eclipse, while interesting, is not the memorable event a total eclipse is.

You've probably figured out that there are only small portions of the Earth's surface over which the Moon's shadow passes during a total or annular solar eclipse. This *eclipse track* begins at sunrise and moves east across the Earth, ending at sunset, almost half the world away. The track may be more than a hundred miles wide or as little as a few feet. The latter case describes the boundary between a total and an annular eclipse. The duration of annularity or totality is greatest at the middle of the track. The partial phase of a solar eclipse can be observed from wide areas of the Earth's surface to either side of the eclipse track.

At the risk of warning you too often, I am compelled to remind you again that *viewing an eclipse with the unprotected eye when the Moon is NOT covering ALL the Sun is no less dangerous than viewing the Sun unsafely at other times.* It's only safe to view a solar eclipse without proper protection when the Moon is *completely* covering the Sun.

One final bit of eclipse vocabulary: *first contact* occurs the moment the Moon's disc first "touches" the Sun's disc. *Second contact* occurs when the Moon has just covered the Sun "on the way in." *Third contact* occurs when the Moon has last covered the Sun "on the way out." *Fourth contact* occurs when the Moon is "last touching" the Moon "on the way out."

3.3.3.2   How Lunar Eclipses Occur

Depending on the relative positions of the Sun, Moon, and Earth, the Moon passes through the Earth's shadow either half a lunation before a solar eclipse or half a lunation after. As you can see from Figure 5, the Earth's shadow is divided into two zones. The *penumbra* or partial shadow surrounds the *umbra* or full shadow.

When the Moon passes only through the penumbra or only partially into it, we have a *penumbral eclipse*. Since the Moon is still in direct sunlight, sometimes a penumbral eclipse can go by almost unnoticed.

A *partial lunar eclipse* occurs when part of the Moon passes through the umbra of the Earth's shadow. If all of the Moon passes through the umbra, a *total lunar eclipse* occurs. When either a partial or total lunar eclipse takes place, you can see incontrovertible proof that the Earth is circular: Its

Figure 5: Lunar and Solar Eclipse*

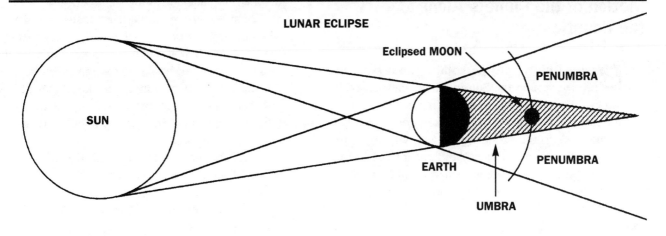

LUNAR ECLIPSE

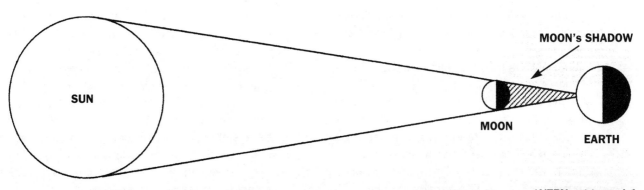

SOLAR ECLIPSE

*VERY not-to-scale!

shadow can be seen passing across the face of the Moon and the shadow is definitely part of a circle.

One final point about lunar eclipses: If you can see the Moon while it is passing through the Earth's shadow, you can see a lunar eclipse. You don't have to be at any special place, with this obvious requirement: The eclipsed Moon MUST be above the horizon to be seen!

3.3.3.3 Eclipse Seasons Happen About Twice Each Year

Earlier, I mentioned that the Moon crosses the ecliptic at two nodes and that the line of those nodes connecting these two crossings precesses westward, taking about 18.6 years to complete one revolution. This concept is fundamental to your understanding of when eclipses occur.

Each new moon, the Moon passes very near the Sun in the sky, but we can't see it because of the Sun's brilliance. Each full moon, the Moon passes very near the Earth's shadow, but we don't see the Earth's shadow unless the Moon passes through it.

Here's the key element: A solar eclipse can only occur when the Moon is new and near one of the two nodes. The same is true of lunar eclipses, except they occur either at the full moon just half a lunation before or after a solar eclipse. Away from the nodes, the Moon passes either too far north or south of the Sun to eclipse it. The same is true for lunar eclipses.

So, solar eclipses occur at new moon at intervals of six lunar months or every six lunations, which means there are a pair of eclipse seasons each year, at approximate intervals of 6 times 29.53 or 177.2 days. Twice that number is 354.4 days, or about ten days less than one year, so eclipse seasons occur about ten days earlier each year. In some years—for example, 1992—there was an eclipse season in very early January, one in late June, and a third in late December. On the average, though, there are only two eclipse seasons a year. (Just because there is an eclipse season doesn't guarantee a total eclipse will occur. It only means some kind of eclipse is likely.)

3.4

Motion of the Planets Along the Ecliptic

Besides Pluto, all the solar system planets revolve around the Sun in planes inclined only very slightly to the plane of Earth's orbit. That means all the planets we can see easily will be found close to the ecliptic in the sky. Where you're likely to find them and how they appear to move on the Celestial Sphere are governed by whether they orbit closer or farther from the Sun than Earth.

3.4.1

The Inferior Planets—Mercury and Venus

Planets orbiting closer to the Sun than Earth are called *inferior* planets, which is a technical term for "orbits closer than Earth to the Sun." These planets can do a few things that others, the *superior* planets, cannot.

1. Inferior planets pass through a similar full range of phases as Earth's Moon.
2. Mercury and Venus don't do it very often, but they can pass exactly between Earth and the Sun. When that happens, we can see a tiny circular dark spot pass slowly across the solar disc. *However, we see this ONLY while viewing the Sun's disc PROPERLY. Never view the Sun without understanding the SAFE ways to do it!*

3.4.1.1   Aspects and Phases of Inferior Planets

As we see it from Earth, as an inferior planet revolves around the Sun, it reaches various positions relative to the Sun. These positions are known as the *aspects* of the inferior planet.

Look at Figure 6 and I'll take you through the aspects and phases of the two inferior planets.

This diagram assumes that we're viewing the inner Solar System from a point far above the North Pole of the Sun. This perspective might lead you to believe all the objects are lying in the same plane. That's not quite true, but it is close.

When using this diagram to study the various aspects of the Earth, Sun, and an inferior planet, remember the diagram is not meant to show you a dynamic or moving situation. It's really a collection of snapshots to illustrate aspects of the three objects.

(Although I defined it earlier when discussing the phases of the Moon, here's the definition of the *synodic period* of an inferior planet: the period that

elapses between successive times when Earth and the inferior planet are at the same aspect.)

Consider *inferior conjunction*, position 1. That's when the inferior planet is lined up exactly between the Earth and Sun. If the lineup is really precise, the inferior planet can be seen crossing the solar disc. For most inferior conjunctions, the planet passes either north or south of the disc. And, of course, at inferior conjunction we can't see the planet because its dark side faces Earth.

Later, the faster–moving inferior planet has "pulled ahead" of Earth, and may be found in the morning sky west of the Sun, where it shows a crescent phase just like the Moon. The inferior planet is at its brightest before it reaches its greatest angular separation from the Sun, mainly because it is so close.

When the planet is at its greatest angular separation from the Sun, we can see the familiar quarter-moon phase. This stage of the planet's travels is called the *greatest western elongation*, position 2.

As the planet continues to recede from us, its phase continues to wax toward full. The planet's magnitude diminishes, even though we can see more and more of its sunlit surface each day. That's because the planet is receding from us so fast that the extra light from the larger proportion of the surface we can see is overwhelmed because the planet is farther and farther away.

When the planet is opposite the Earth from the Sun, it has reached *superior conjunction*, position 3. Because it is either behind the Sun or only a bit north or south of its disc, we can't usually see an inferior planet at its superior conjunction.

After passing superior conjunction, the planet passes into the evening sky, where its phases are the reverse of what it showed us in the morning sky. When it once again shows half-phase, the inferior planet has reached *greatest eastern elongation* from the Sun, position 4. At the end of one synodic period, the inferior planet is again lined up between the Earth and Sun; then, the cycle repeats.

3.4.1.2   Finding Mercury in the Sky

Because it is the planet closest to the Sun, Mercury is also the fastest moving, and the most difficult to observe. Mercury orbits the Sun in only 88 days, and its synodic period is 116 days, which means we can see several apparitions of Mercury every year.

Not only is Mercury's orbit closest to the Sun, but its orbit is also one of the most eccentric of all the planets, or the most non-circular. Consequently, Mercury's greatest elongation from the Sun can be as little as 18 degrees and as much as 28 degrees.

In the early spring the ecliptic makes a fairly steep angle with the western horizon just after sunset. In that season, Mercury's elongation from the Sun translates to its maximum altitude above the horizon just after sunset. In general, then, the best evening apparition of Mercury each year will occur in early spring. In early fall, the same geometry applies to viewing Mercury just before sunrise.

One final word on finding Mercury: This planet is not visible for long after sunset or before sunrise, regardless of the time of year you view. So, just like the Boy Scout, be prepared.

3.4.1.3 Finding Venus in the Sky

Venus is farther from the Sun than Mercury, so its orbital period is longer—225 days. Its synodic period is 584 days, which indicates that apparitions of Venus last much longer than Mercury. Not only can they last longer, but Venus can be found as far as 48 degrees east or west of the Sun along the ecliptic.

NOTE: When Venus appears in either the evening or morning sky, it's usually BRIGHT. In fact, Venus is misidentified so often (because of its extreme brightness) that it's usually the number one cause of UFO reports.

3.4.2 Superior Planets—All the Rest

The planets that orbit beyond Earth are called the superior planets. This grouping is only appropriate when we're discussing how the planets move along the ecliptic. As you'll learn in Chapter 7, Mars bears little physical similarity to Jupiter, Saturn, Uranus, and Neptune. Pluto is a real oddball.

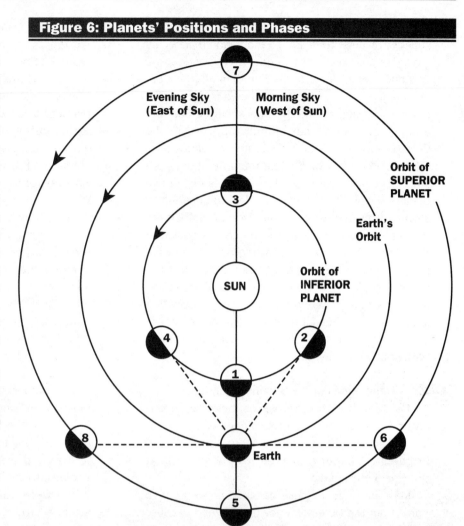

Figure 6: Planets' Positions and Phases

3.4.2.1 Aspects and Phases of the Superior Planets

Stuck here on Earth, we can't get behind the superior planets, so we can't see much of their dark sides. We can see only a slightly gibbous phase when a superior planet is at *quadrature*, the term for "the superior planet is 90 degrees from the Sun."

A superior planet is at quadrature when it's at position 6 or 8 on Figure 6. For superior planets, the synodic period is quite long for a planet close to Earth, shrinking as we go outward from planet to planet. Rather than devote many wordy paragraphs to this, just read this table:

Planet	Orbital Period	Synodic Period
Mars	687 days	780 days
Jupiter	11.9 years	399 days
Saturn	29.5 years	378 days
Uranus	84.0 years	370 days
Neptune	164.8 years	367 days
Pluto	247.7 years	367 days

Figure 6 shows the aspects and phases of the superior planets. There is a phenomenon that occurs around opposition that isn't explained by the diagram.

Earth, being closer to the Sun than all the superior planets, orbits faster than they do. Thus, when Earth is moving into position between the Sun and a superior planet, that planet appears (on the Celestial Sphere) to be moving toward opposition. On two occasions, one before and one after opposition, the combination of the motions of the two planets results in the superior planet appearing to stop and reverse its direction. Around opposition, the superior planet appears to describe a backward **S** or loop in the sky, depending on the relative inclination of the planes of the Earth's and the planet's orbits. Astronomers call this a retrograde loop. Inferior planets play "here we go loop-the-loop," too, but they all do it at the time of inferior conjunction, when we can't watch, another example of their inherent shyness.

3.4.2.2   Oppositions of the Superior Planets

The best time to observe a superior planet is around opposition when the planet is closest to Earth, so its disc appears at its largest and brightest. A superior planet at opposition is at position 5 on Figure 6.

Mars is a special case because it is closer to Earth than the other superior planets and because its orbit is more eccentric than other superior planet orbits. So there are oppositions, and then there are oppositions.

A *perihelic opposition* occurs when Mars is near its closest point in its orbit, its perhelion. Then Mars is only about 35,000,000 miles from Earth and its disc has an angular diameter of as much as 24″. At those times, Mars can appear in a small telescope as big as the Moon does to our unaided eye.

Unfortunately, these favorable oppositions of Mars occur only every 15 or 17 years. There was one in 1956, another in 1971, and a pair in 1986 and 1988. The next favorable opposition will be in 2003.

Oppositions of Mars that occur when Mars is not near its perhelion are much less favorable viewing than perhelic oppositions. The apparent diameter of Mars's disc can be in the range of 13″ to 16″, which doesn't allow nearly so much detail to be seen.

3.4.3

Constellations of the Zodiac

The twelve constellations lying along the ecliptic make up the Zodiac. You can also think of the Zodiac as a band running along the ecliptic that is 16 degrees wide, in which the planets may be thought to be confined.

Since there are twelve months in a year and twelve constellations in the Zodiac, you might be tempted to assume the Sun spends one month in each. Not true. The constellations of the Zodiac vary quite a bit in their frontage along the ecliptic.

The Zodiac is a leftover from the time long ago when astronomers were astrologers and vice versa. When astrology was developed as a "science," people thought that the positions of the Sun and planets along the ecliptic at the time of a person's birth had great influence on the course of that person's life, including influencing the person to take on the characteristics of the mythological figure represented by the constellation. Thus, a person born when the Sun was in the constellation Taurus (the Bull) would be stubborn as a bull (and, possibly, "bullish on America").

However, some constellations of the Zodiac have more frontage on the ecliptic than others do. This means that Taurus, which has a long frontage, was influencing the births of more people than Cancer, and producing too many really stubborn people.

To avoid that problem, astrologers divided the Zodiac into twelve "signs," each covering 30 degrees. If you were born when the Sun was in the area assigned to a particular sign, your life would be influenced by the characteristics of that sign. If you were born on January 26, you were born under the sign of Aquarius the Water Bearer and would, presumably, spend your days with a bottle of water strapped to your back.

I've obviously had a bit of fun with astrology, which is no longer thought to be a science. However, if you wish to read your daily horoscope for your own amusement, don't let me stop you. For your sake, I wish you the following: That you'll lead a safe life every day, not just on those days when your horoscope tells you to be careful....That you'll be careful with your money every day, not just when your horoscope tells you to....That you'll be a caring, loving person every day, not just when your horoscope tells you to pay attention to romance....Get the picture?

A final word about astrology: The Astronomical Writers' Union requires me to point out that, because of the precession of the Earth's axis, the Sun is NOT in the constellation represented by its astrological sign that the "science" says it should be. Therefore, if you were born under the sign of Taurus the Bull and have led a stubborn life, be warned that the Sun was really in Aries the Ram when you were born. Maybe you shouldn't be stubborn, but butting heads with other Aries. Or, maybe you should just live your life being you and not who your horoscope says you should be.

Mini-almanac (and Almanacs in General)

Since they're moving hither and yon (mostly yon), the positions of the planets cannot be plotted on the maps, so I've produced a mini-almanac that will tell you where to find the planets for each month during the years 1993–2000. You can also find on what date the Moon will be at what phase, and when and where solar and lunar eclipses will occur. Section 3.6 tells you how to use this mini-almanac.

There are a number of almanacs published each year and most of them feature astronomical tables that provide much more information than I can in this book. If you want a simple almanac, I recommend the *Old Farmer's Almanac*, which includes weather predictions for the entire year that seem slightly more accurate than horoscopes. Other features you might find useful are tables of sunrise and sunset, moonrise and moonset.

There are two publications you should also consider: They are the *Astronomical Calendar*, published by Furman University Press, and *The Observer's Handbook*, published by the Royal Astronomical Society of Canada. Both provide a wealth of information. The *Calendar* is a large format book with lavish graphics, while the *Handbook* is smaller, but provides a wider variety of material. I recommend both, but seem to use the *Handbook* a bit more often, because I can carry it with me.

Both *Astronomy* and *Sky & Telescope* feature extensive almanac-style information each month. *Astronomy*, for which I've been a columnist since 1988, is intended for the beginning-to-intermediate amateur astronomer, while *Sky & Telescope* is intended for more hardcore viewers. I read and recommend both.

Besides providing more material than I can in this book, these two magazines will also keep you up to date on the appearance of a new comet or the sudden brightening of a star. Since my crystal ball is as cloudy as a horoscope, I can't tell you when events like those will occur.

Using the Maps and Mini-almanac to Find the Planets

You'll find the mini-almanac in the Appendix in the back of this book. I've tried to design the mini-almanac so it provides the most useful information possible and still is easy to use.

Using the Mini-almanac

There's one block in the mini-almanac for each year from 1993 to 2000. Don't make the obvious mistake of looking at the block for 1994 when it's really 1993, or you're not likely to find what you seek.

The layout of each block is simple. The months of the year are in the first column. The second tells you the dates when the Moon will be in what phase. Positions of the bright planets are found in the third column. Finally, details of the eclipses that happen that year are found in the last column.

To Determine the Phase of the Moon

Let's go through the month of May 1993: We find the Moon was full on the 6th, because the number 6 appears in the column marked "full," in the row marked "May." On the 13th, the Moon was at third-quarter phase, and new on the 21st. Jump down one line, still in May, and note that the Moon showed its first-quarter phase on the 28th.

If you wish to observe a particular feature on the Moon, you may find it's best seen when the Moon is a certain number of days old. You can use the mini-almanac to determine that, too. Suppose you wanted to observe the Moon in May 1993, when the Moon was five days old. Here's how:

First, new moon was when the Moon was zero days old. The Moon was five days old approximately five days after new moon. We already found that the Moon was new on May 21, so, adding 5 to 21 gives us 26. The Moon was five days old on the 26th of May.

In May, the Sun was eclipsed at new moon. We know that because there's a capital E next to the date, and because solar eclipses can only occur at new moon. Look in the fourth column under "Eclipse Details," and find that there was a partial solar eclipse on the 21st, and that observers in North America and northern Europe were able to see it. (To find out more details on this eclipse, consult one of the publications listed in Section 3.5.)

3.6.1.2   To Find the Inferior Planets

Since the inferior planets, Mercury and Venus, never stray far from the Sun, I've only indicated whether they are evening or morning objects in the mini-almanac. The first column under "Bright Planet Positions" is labeled M for Mercury, the second is labeled V for Venus.

Again taking May 1993 as an example, we find Venus was a morning object then, because we found an M where May's row and Venus's column intersect. Mercury is not visible, because it is in conjunction with the Sun.

Some words of caution: As I pointed out in Section 3.4.1.2, although it makes several morning and evening apparitions each year, Mercury only makes one good morning apparition and one good evening apparition each year. The good morning apparition occurs in early fall and the good evening apparition in early spring.

3.6.1.3   To Find the Superior Planets

Since Mars, Jupiter, and Saturn can be found anywhere along the ecliptic, not just near the Sun, I've listed the constellation in which the superior planets may be found each month. There are only twelve, since the planets only move within the constellations of the Zodiac. The constellations, and their abbreviations:

Aquarius: AQR	Aries: ARI
Cancer: CAN	Capricornus: CAP
Gemini: GEM	Libra: LIB
Leo: LEO	Pisces: PSC
Scorpius: SCO	Sagittarius: SGR
Taurus: TAU	Virgo: VIR

Uranus and Neptune are so bright they're easy to find. However, they don't move very far on the Celestial Sphere from year to year. Both planets are near each other in the sky, moving from eastern Sagittarius into Capricornus. To find their positions, consult *Astronomy* or *Sky and Telescope*.

3.6.1.4   To Find When an Eclipse Will Occur

As you read in Section 3.3.3, solar eclipses occur only at new moon, while lunar eclipses only occur when the Moon is full. To denote those dates when an eclipse will occur, there is a capital E just to the right of the date of the phase of the Moon. If you notice that either a new-moon or a full-moon date is accompanied by the letter E, just look at the last column of the table, where you will find details of the eclipse.

3.6.2

Identifying a Mystery Planet

All constellations of the Zodiac are marked on the ALL-Sky maps, as well as on the ZOOM maps. If you notice a bright star-like object in a constellation of the Zodiac, and it obviously doesn't fit, it's very likely one of the planets. Look at the mini-almanac page for the year you're viewing and note which planet is predicted to be in that constellation. And voilà!

3.6.3

Finding a Particular Planet

3.6.3.1   The Inferior Planets

If you wish to find one of the inferior planets, look at the mini-almanac page for the year in which you're viewing. You will note that the planet is either in the morning M or the evening E. Then find the ALL-Sky map closest to the time you wish to observe the planet. This will usually be the map for the time one hour after sunset, or one hour before sunrise.

If the planet you seek is an evening object, note where the ecliptic meets the western horizon on the map. Find the due-north point using any of the methods outlined in Section 2.5. Stand with your arms spread straight out from your sides, your right hand pointing north. When you look straight ahead, you'll be facing due west. You'll be able to trace the approximate point where the ecliptic meets the western horizon. In your mind's eye, trace the ecliptic from the horizon, identifying the constellations of the Zodiac and other bright objects that are near the horizon. The bright object near the ecliptic—the one that doesn't fit—is the planet you seek.

Finding a morning object is no different in principle than finding one in the evening, except you will be seeking it on the eastern horizon, and when finding the due-east point, you will stand with your left arm pointing north.

3.6.3.2   The Superior Planets

When finding one of the superior planets you will employ a different strategy, because they are not near the Sun in the sky as often as the inferior planets. Using the mini-almanac, find the constellation in which the planet is predicted to be. Find the ALL-Sky map for the time you wish to observe the planet and locate the constellation. When you've found it you will likely note that there is a bright star that doesn't fit the pattern of the constellation. That's the planet you seek.

CHAPTER 4

Techniques for Viewing the Sky

Many people simply enjoy just looking at the sky, knowing where the constellations are. Or they follow the planets moving through the sky. Others use binoculars or modest telescopes. Some skywatchers spend thousands, even tens of thousands of dollars on elaborate equipment, sometimes housing it in compact observatories away from city lights. Unless you're preserving images of the sky, merely observing it comes down to looking at it with your eyes, either aided by some helpful optics or unaided. Before we discuss your eyes, though, let's consider what goes into your eye—light.

4.1

A Very Short Primer on Light

Every object in the sky either emits, reflects, or absorbs light. So, what is light? The short answer is that we really don't know, and we may never. However, scientists have developed a model that explains most of the phenomena associated with light. Fortunately, when you build this model of light, you won't get "airplane glue" all over your fingertips. However, there are a few other details involving light that are quite sticky!

4.1.1
The (Lickety-Split) Photon

What we see as light is made up of very energetic particles called photons. These guys have no mass, but they make up for this by going extremely fast, about 186,000 miles per second. In fact, they can only go that speed—no faster, no slower. (If they're not in a vacuum, photons have to slow down, but they're still going nearly as fast as in a vacuum.)

4.1.2
The Sociology of Photons—Light Waves

Although many experiments have proven that light is made up of particles, other experiments have proven, just as conclusively, that light behaves as continuous waves of energy. In fact, some experiments seem to indicate that light changes its behavior from a wave to a particle phenomenon if the nature of an experiment is changed half-way through! This leads a few scientists to wonder if the humans are experimenting on the photons, or the photons experimenting on the humans!

A branch of physics called quantum mechanics attempts to describe how light can be both discrete particles and a continuous wave. Quantum mechanics recognizes that individual photons are very social particles that almost always bow to peer pressure and "go along with the wave," much as individual baseball fans seem to participate in "the wave" during the more boring parts of a game. So, quantum mechanics can be described as the "Sociology of Photons."

Going any further into quantum mechanics just eats up pages (and trees!) without helping you understand what you need to know to enjoy looking at the sky. Once I've told you how your eye and other basic optical instruments work, I'll tell you why the wave-nature of light is important.

4.1.3
The Electromagnetic Spectrum and the Color of Light

The length of a wave of light is what determines the color of light we see. This wavelength is also inversely proportional to the energy of the photons making up the wave. Low-energy photons have long wavelengths, and high-energy photons have short wavelengths. (Since light cannot travel at any speed but the magic 186,000 miles per second, it changes wavelength when it either loses or gains energy, instead of changing speed as other particles do.)

Our eyes can see only a small portion of the vast array of photons that are careening around the Universe. The full range of energies that photons can have is called the electromagnetic spectrum. Here's a table of the EM spectrum:

The Electromagnetic Spectrum

Type of EM Radiation	Wavelength		
Gamma Rays (highest energy)	0.0000000001	or	10^{-10}cm
X-Rays	0.00000001	or	10^{-8}cm
Ultraviolet Light	0.000001	or	10^{-6}cm
Visible Light (medium energy)	0.0001	or	10^{-4}cm
Infrared Light	0.01	or	10^{-2}cm
Microwaves	1.0	or	10^{0}cm
Radio (low energy)	100.0	or	10^{2}cm
Radio (ultra low E)	10000.0	or	10^{4}cm

Before discussing the importance of the table above, I'll take a moment to discuss the way of writing very large and small numbers I've used in the third column of the table. It's simple to write very large numbers in the way we all learned in grade school. If you want to write "a million," you just write a 1 followed by 6 zeroes. A million-million becomes a 1 followed by 12 zeroes. (Presumably, a jillion becomes a 1 followed by as many zeroes as you can write.) You can also write very small decimal fractions by just being willing to write a lot of zeroes.

But, simple isn't always easy. The system I've used in the third column is called scientific notation. It allows you to write the number 1,000,000 as 10^6, or the fraction 0.000001 as 10^{-6}. If you want to write the number of atoms in one gram of matter, you can either write

60,200,000,000,000,000,000,000 or 6.02 x 10^{23}.

This is said, "six point zero two times 10 to the twenty-third power." To get "ten to the twenty-third power" you would have to multiply 10 times itself 23 times.

So, depending on how energetic a photon is, it "hangs out" as part of a long-wavelength radio wave, or really lives life in the "fast lane" as part of a gamma-ray wave, or anything in between. Fortunately, the Earth's atmosphere protects us from the more energetic photons than those we see with our eyes, although someone who's got a good sunburn from ultraviolet rays may disagree!

4.1.4

The Spectrum of Light Provides Clues to What's Going On

You've probably seen what happens when a pot of salted water boils over on the stove. The water

contacts the flame, and you see a very yellow light. Salt is a compound made up of one atom of Sodium and one of Chlorine. That yellow light is characteristic of Sodium being heated. It's Sodium's "signature."

Figure 7 shows how a prism disperses light into the spectrum. However, I'll bet your first view of the spectrum was when you first saw a rainbow. Rainbows are caused when light is bent, or refracted, through raindrops. Dispersion occurs because red light is refracted less than green light, which is refracted less than blue, and so on....

Figure 7 shows a simple spectroscope. Such an instrument allows an astronomer to break up light into its various colors. Each substance that's present in an object that's emitting or absorbing light has a characteristic spectrum that allows the astronomer to determine what's there by looking at the spectrum. Once she knows what's there, the astronomer can put together a theory that explains what's happening to cause the light in the first place.

There are three types of spectra:

1. Bright line spectra occur when a gas is heated. A fluorescent light seen through a spectroscope shows discrete spectral lines, because a gas inside the tube is heated. Each line is the image of the slit seen at one discrete wavelength. Each wavelength corresponds to a specific energy level of the electrons surrounding the nucleus of the atoms that are emitting the light. Only a few distinct levels are allowed, so only a few lines are seen.

2. A continuous spectrum is caused by a solid substance being heated.

3. A dark line spectrum is caused when a cool gas is between the observer and a source of a continuous spectrum. The electrons of the atoms making up the gas absorb photons of light at only a few allowed wavelengths, thus removing those wavelengths from the spectrum. We see a dark "image" of the slit in the spectroscope.

The spectra of the Sun and of other stars are a complex mix of bright and dark line spectra. However, much of the solar spectrum appears to be continuous. Is the Sun solid? No! Rather, its spectrum is made up of many bright line spectra that are all being emitted under a wide variety of conditions. They all superimpose upon one another

Figure 7: A Simple Spectroscope

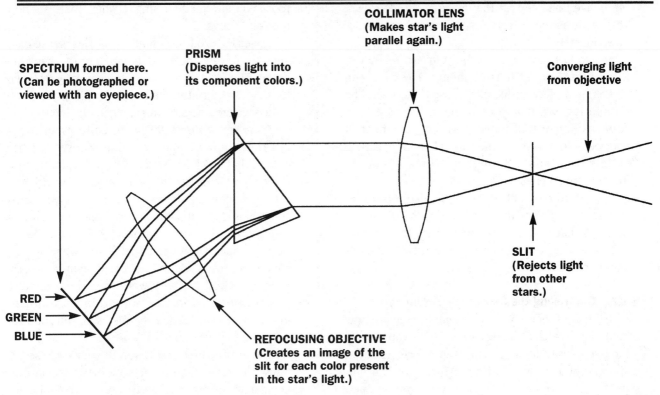

SPECTRUM formed here.
(Can be photographed or viewed with an eyepiece.)

PRISM
(Disperses light into its component colors.)

COLLIMATOR LENS
(Makes star's light parallel again.)

Converging light from objective

SLIT
(Rejects light from other stars.)

RED
GREEN
BLUE

REFOCUSING OBJECTIVE
(Creates an image of the slit for each color present in the star's light.)

to simulate a continuous spectrum. The bright lines are caused by incredibly hot gases, and the dark lines by cooler gases absorbing light at those wavelengths.

As I mentioned, studying the spectra of the various objects in the sky is one of the fundamental ways astronomers learn what's going on. If you know a bit about spectroscopy, you can better understand what you're seeing, too.

Over the years there have always been small spectroscopes available for amateur telescopes. The user just points his telescope at a bright star, and then removes the normal eyepiece and inserts the spectroscope. Instead of the sharp image of the star seen through the eyepiece he sees the spectrum of the star.

I have such a stellar spectroscope. When I've shown people the spectra of stars through my 20-inch telescope, they've been positively enthralled by the view. We've looked back and forth between stars of different color, and were easily able to notice the difference in the spectra of the two.

Stellar spectroscopes are neither inexpensive nor popular. I suspect it's because they work well with very bright stars, and only really well with fairly large telescopes. However, they certainly add a new dimension to what can be seen through a telescope.

4.2

About Your Eyes

Your eyes are amazing instruments. Every day, an enormous amount of information pours through them and is processed by your brain. Before I get into how they work, here's yet another warning: Never, *never look at the Sun with your eyes unprotected*. Your eyes are just too valuable to endanger.

4.2.1

How Your Eyes Work

Technically, your eye is a sophisticated camera. It consists of a lens, iris, sensor array, and an image-processing system. I'm not going to go into great detail describing the eye, but will touch only on those aspects of its performance that apply to looking at the sky.

4.2.1.1   The Lens in Your Eye

The lens in your eye is similar to the lens in an ordinary magnifying glass. If you have a magnifier, take it into a room with a window at one end. Hold your magnifier up to the wall opposite the one with the window. Move the magnifier in and out until you see a sharp image on the wall. (You've just focused the image.) Notice that the image is upside down. That's why astronomical telescopes show an upside-down image.

Instead of having one lens, your eye really has three parts, called lens elements. The first, the cornea, is a piece of literally transparent flesh. The cornea faces out and contains a liquid called aqueous humor that forms the second element of your eye's lens system. The third element is called, simply, the lens, although all three elements function together as a lens.

This third element, the lens, is also a piece of transparent flesh. It's surrounded by a ring of muscles that either squeeze the lens to focus on nearer objects or relax so you can focus on more distant objects.

4.2.1.2 Controlling the Sensitivity of Your Eye

It's the job of the iris diaphragm to regulate how much light reaches the sensor array in your eye. If there is a lot of light present, your iris closes to an aperture of as little as 0.5 mm. (There are 25.4 mm to one inch.) When it's very dark, your eye can open to an aperture of as much as 7 mm, a bit over 0.25 inch. When your eye is completely opened, it can allow about 200 times as much light as when it's closed to its minimum aperture. (That's the ratio of the areas of the 7-mm aperture to the 0.5-mm aperture. Call this the range of brightness your iris can control.)

This range of brightness is only about 200 to 1, yet the range of brightness your eyes can handle is approximately 10,000 to 1. The rest of the range your eye can handle is provided by chemicals that are decomposed in the presence of bright light and recomposed in very dim light. That's why it takes so long for your eye to become adapted to very low-light conditions. Although your iris can open up relatively quickly, the chemicals responsible for your low-light sensitivity take as long as an hour to recompose completely. The light-sensitivity-regulating chemicals are found in the sensor array of your eye.

4.2.1.3 The Sensor Array in Your Eye

How your sensor array works, and how well you understand how it works, have a lot to do with how well you'll see objects in the sky. Your array is called the *retina*. At its center is the *fovea*, which is filled with sensors known as *cones*. Another ring of sensors on the periphery of the eye is filled with sensors

called *rods*. Between the fovea and the outer ring of rods is an intermediate ring in which there are both rods and cones.

Cones do two things really well: They sense color and they're very good at separating or resolving fine detail. That's why we always roll our eyes so we're looking straight at something we want to see well. But cones are not good at seeing in dim light

Rods sense dim light much better than cones, but they flunk at sensing color or resolving fine detail. That's O.K., because they're located around the edge of your array. Their job is to provide you with a wide-angle view, even at night. If you sense something dangerous in your peripheral vision, you can always roll your eyes toward it for examination by your cones. (It appears that we evolved with this characteristic as a survival mechanism. Since the rods don't resolve very well, they don't send as much information to your brain, minimizing information overload. They just send information essential to your survival.)

The area of high resolution in your eye's sensor array, the fovea, can resolve objects that have an angular size of approximately 1'. Since the full moon is approximately 30' in apparent diameter, you should be able to see objects a thirtieth of the full moon's diameter.

There's a catch to this 1' business, though. Your eye can only resolve that small an object if the light level is fairly high and if the contrast between the objects to be resolved is fairly high. If the contrast drops or if the overall light level diminishes, the ability of your eye to resolve fine detail goes down, too.

4.2.1.4 Your Eye-Brain Combination

When you view very dim objects, it appears that your eye-brain combination groups together a number of rods and only registers that something is there if all or most of the sensors in a group are triggered by faint light. The dimmer the object you wish to view, the bigger it must be for your eye-brain combination to be able to notice it.

Your brain is a powerful image-processing system. An example is that the lens in your eye presents your sensor array with an upside-down image, as any lens must. Your brain inverts it for you. Your brain also tends to correct other defects in the image your sensor array receives from your lens.

Furthermore, your eye-brain combination can be trained to see things that are very dim or to see very subtle differences in light level or color.

4.2.2
Getting the Most Out of Your Eye

Many amateur astronomers spend big bucks on fancy equipment for viewing the sky, then fail to get as much out of their equipment as they could. The reason: They don't pay enough attention to maximizing the performance of their eyes.

4.2.2.1 Put in Your Time

There are several ways to use your knowledge of how your eye-brain combination works to maximize what you see in the sky. The most obvious: The more you look, the more you'll see. If you only look at the sky a few minutes each month, you're likely to remain relatively unimpressed with your view. However, if you "put in your time," you'll be surprised at how much more you'll see.

This principle may apply even to viewing certain kinds of objects. So, even if you've observed the Moon many times, you may find you'll have to retrain your eye-brain combination all over again if you take a long hiatus from viewing. The solution? Observe as many different kinds of objects as often as you can. It IS addicting, but it's also GOOD for you!

4.2.2.2 Get Your Eyes Adapted to the Dark, Then Keep Them That Way

Your ability to view dim objects can be impaired if you expose your eyes to very bright light during the day before an evening observing session. Wearing a pair of very dark sunglasses for the last half of the day will prevent that. I've often worn my sunglasses until well after sunset, because they seem to hasten the time it takes my eyes to open up to their maximum.

If you need a light to illuminate a star map or for other reasons, make it a very dim red light. Red is the color of choice because it doesn't seem to cause your eye to open in response to light the way other colors do. In Chapter 2, I told you how to make an astronomer's flashlight.

If you don't want to make one, you can buy accessory red filters for some brands of flashlights. However, they tend to be too bright for my tastes. There is one type of flashlight that employs a light-emitting diode to produce a very red light. It also features an adjustable light level. Whatever light you use, don't point it at other observers' eyes. Keep it pointed at your maps or equipment.

If you must go inside where there are bright lights after your eyes have become adapted to the dark, you can avoid losing your dark adaption by wearing a pair of very dark sunglasses or by wearing a pair of very dark red goggles. Several types of goggles are available.

4.2.2.3 Looking Out of the Corner of Your Eye

Remember, I already pointed out that the rods near the periphery of your eye's sensor array are much more sensitive to dim light than the cones near the center. You can use that fact to see dim objects. Instead of looking straight at an object, look at it out of the corner of your eye. You'll find dim objects seem to brighten up when you do that. (By the way, the technical term for this is "averted vision.")

4.2.2.4 Move Your View Across the Field

Suppose you're looking for a very dim object like the end of a comet's tail. You can't see it, even with averted vision, presumably because it's not big enough to trigger enough rods to convince your brain something is there. If you do something to move your view rapidly across the field, though, you suddenly notice it pop into view.

The explanation for this is not simple, but it appears to be a result of sophisticated image-processing capabilities we humans developed while we evolved in an environment in which survival depended on detecting very dim objects in our peripheral vision. Those of our forebears who didn't have this ability, and who weren't constantly shifting their vision all over the place, likely didn't survive long enough to procreate. They certainly didn't have your view of a dim comet's tail in mind, because just surviving until sunup was paramount for them.

If you're observing without an optical aid, it's hard to scan your view across a particular area of the sky, because your eyes tend to lock onto a bright star and try to follow it. One trick is to use your hands to limit the view drastically, then rock your head up and down or back and forth. You'll find your eyes will not follow this restricted field.

If you're using binoculars, you can just rapidly pan them across the desired field to try to make the dim object reveal itself. Giving a telescope a gentle shove is often the best way to achieve the desired effect. Ironically, the sturdier the telescope mount, the less likely you will be able to use this technique. However, I don't recommend a flimsy mount for your telescope just so you can rock it easily!

4.2.2.5 Take Good Care of Yourself

I doubt if you need me to tell you that, but if you're tired, your eyes won't work as well as when you're well rested. That's true. The best thing you can do for your ability to see things in the sky is to get enough rest before you go out to view. If you can get a catnap for a few minutes after dinner, you'll probably find you'll see more, because you'll feel like seeing more.

Your choice of food can help you see the sky better, too. No, I'm not going to tell you to go on a diet of carrots! However, eating a relatively high-protein evening meal is better than loading up on carbohydrates, because your system will take a long time to digest the protein and will be supplying your body with a steady flow of energy for a long time into the evening. When my vision is starting to slow down in the late evening, I find a snack that combines carbohydrates and protein gives me both a quick boost and some long-term energy.

4.3

How a Telescope Works

Entire books have been devoted to telescopes, so the first thing you need to know is that you need to know more than I can tell you in the space we have here. However, I've tried to give you some basic principles you can apply to making decisions about what kind of telescope might be right for you. What you learn here will also make it easier for you to use your telescope.

4.3.1

Forming an Image

If you've tried out your magnifying glass to form an image, you have a good idea how the first half of a telescope works. Forming an image of the sky is extremely important. No amount of money spent on all the other accessories you can add to a telescope will correct an image badly formed by a poor lens or mirror. (An exception is the incredible amount of our money that had to be spent on computer-processing the images from the Hubble Space Telescope to partially compensate for its badly formed mirror.)

In the following sections, you'll read a lot of material on how an image is formed. However, it all boils down to this: The lens or mirror that forms the image must do two things—intercept as much light coming from the stars as possible and form that light into as sharp and true an image of the stars as possible.

4.3.1.1   From Prism to Lens

While we were thinking about the nature of light, I brought up the spectroscope, and briefly explained how the prism refracted the various colors of the visual spectrum. Now we'll build on those concepts to learn how a lens can form an image.

Now look at Figure 8. I drew the lenses quite large, and you might be tempted to think that this drawing is a bit wasteful. However, they're larger than those shown in other books in order to make the basic features of a telescope easy to see.

The surfaces of a prism are flat, because it's used only to disperse the light into the colors of the rainbow. A lens uses refraction to form an image. To do that, it has a curve, usually a sphere, that bends the light rays near the center of the lens the least, and those near the edge the most. (By the way, light "rays" don't really exist. They just indicate the direction the waves are going.)

For now, all I want you to get from Figure 8 is to see that the parallel rays from three different stars in the field of view of our telescope are brought to a focus to form an image. They're parallel because the stars are infinitely far away from the lens.

As Figure 8 shows, the lens that forms the image is called the objective lens, or just the objective, because it forms an image of an object. (In this case, the "object" is a group of stars.) The distance between the objective and the image is the focal length. The longer that distance, the farther the images of the stars will be apart, and the bigger the image will be. Since the lens forms an image by refracting the light that passes through it, it is called a refractor objective.

So, we've done what we set out to do. We have an image of a group of stars. And, it appears that the various rays all converge on the same point to form a perfect image of the stars. That would be true, except for a few minor details.

4.3.1.2   Those Minor Details—Aberrations

The first "minor detail" is really a major detail. In order to avoid clutter, I drew Figure 8 as if the rays of light passing through it were of only one wavelength, say, green. It's a great telescope, if all you want to look at is stars that shine in one wavelength of green light! But, remember, when light of many colors is refracted, the light is dispersed into a spectrum. It's the same with lenses. The red image of the stars will be found below the green image in Figure 8, and the blue image will be found closer to the objective lens. Since the size of the image depends on the distance between the lens and the image, the blue image will be smaller than the green, which will be smaller than the red image! This defect of a single-lens objective is called *chromatic aberration*. (An aberration is a defect in an image.) It's not a pretty sight!

To correct this problem, objective lenses are designed with two or more separate lens elements, just as your eye has several elements. If there are two elements, the lens is called an *achromatic lens*. Two distinct wavelengths of light can be brought to the same focus, and the rest are close. To minimize the "leftover" or residual chromatic aberration of an achromatic lens, the ratio of its focal length to its diameter is usually 15 or greater.

Objective lenses featuring three lenses bring three distinct wavelengths to the same focus, and have almost no residual chromatic aberration. These lenses are called *apochromats*. As you might expect, they are more expensive than achromats.

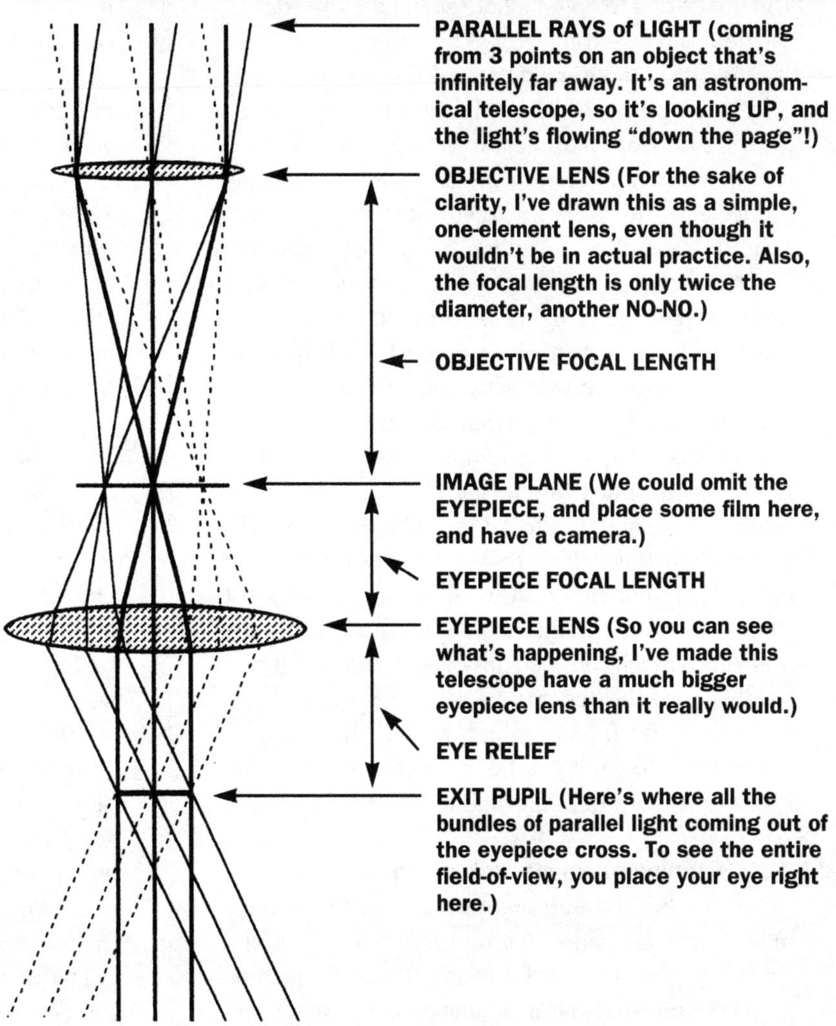

Objective lenses suffer from a number of other aberrations. Rather than take up many words, and as many pages of diagrams explaining them, I'll just list them here:

Curvature of Field: Figure 8 would have you believe that the surface on which the image forms is perfectly flat. That's close, but not strictly true. The "surface of best focus" is usually a sphere of very long radius, curving toward the objective lens.

On-Axis Aberrations: The chief offender of this group of aberrations is *spherical aberration*. This just refers to the fact that many lenses have a slightly different focal length for rays that strike them near their edges than they do for rays striking near the center. The result is that a star image isn't sharp, but is a bit fuzzy.

Off-Axis Aberrations: There are two chief offenders: *Astigmatism* is seen as a star image that looks like a line, and *coma* causes a teardrop-shaped star image. Astigmatism and coma are called off-axis

Figure 8: A Simple Telescope

PARALLEL RAYS of LIGHT (coming from 3 points on an object that's infinitely far away. It's an astronomical telescope, so it's looking UP, and the light's flowing "down the page"!)

OBJECTIVE LENS (For the sake of clarity, I've drawn this as a simple, one-element lens, even though it wouldn't be in actual practice. Also, the focal length is only twice the diameter, another NO-NO.)

OBJECTIVE FOCAL LENGTH

IMAGE PLANE (We could omit the EYEPIECE, and place some film here, and have a camera.)

EYEPIECE FOCAL LENGTH

EYEPIECE LENS (So you can see what's happening, I've made this telescope have a much bigger eyepiece lens than it really would.)

EYE RELIEF

EXIT PUPIL (Here's where all the bundles of parallel light coming out of the eyepiece cross. To see the entire field-of-view, you place your eye right here.)

aberrations because they do not affect the image of a star that's in the very center of an objective's field-of-view ("on-axis"), but progressively affect the images of stars as they are farther and farther toward the edge of the field (farther and farther "off-axis"). Oftentimes, coma and astigmatism are both present, and they combine to spread the image of a star into a misshapen disc, instead of the nice sharp point that's desired.

Distortion: If you were to point an objective at a distant window screen, and then examine the image, you might very well find the screen in focus. However, if the objective suffered from distortion, only the square exactly in the center of the field would look square. The ones next to the center square would look stretched a bit, and the ones farther out would be stretched a little more. The ones near the edge would look more like diamonds than squares. You can often see distortion in the images produced by very "wide-angle" binoculars. The image of a telephone pole seen near the edge of the field will be strongly curved.

Distortion is an aberration of the overall image, and is often ignored by designers of general-purpose optical instruments. However, if you use a pair of binoculars to view the sky, and you pan the binoculars across the sky, distortion will cause the stars to appear to be "rolling" along the surface of a sphere, rather than appearing to move across a flat surface. Many users of binoculars find this annoying, and manufacturers of high-quality binoculars intended for astronomical use are now correcting for distortion in their designs.

No lens design can eliminate every aberration. However, a lens designer can make all of them reasonably small. Sometimes, the designer will make a lens that minimizes aberrations over a wide field, but is not particularly sharp in the center of the field. If the image the lens forms will not be magnified much, the person viewing through it will hardly notice the aberrations. However, if the objective will be used at very high power, the designer might have to limit the field of view to allow the image in the center of the field to be perfect.

4.3.1.3   Forming an Image with a Mirror

Remember the mirror your mother used to have in the bathroom, with a flat mirror on one side and a magnifying mirror on the other? Be truthful, how many of you used the magnifying side to focus the Sun's image on a colony of ants and played "War of the Worlds"? If you did, you were using an objective mirror, no different than the one you might find in a reflector telescope, except for its quality. The mirror objective is concave, and the surface facing the sky is coated with a reflecting metal, usually aluminum.

Reflector objectives usually have the shape of a paraboloid, which is similar to a sphere, only the center is a bit deeper than a sphere would be. Because light is not dispersed when reflected, mirror objectives do not suffer from chromatic aberration. Because they do not have a spherical surface, they do not suffer from spherical aberration. Reflector objectives do suffer from all the other aberrations that plague refractor objectives.

4.3.2
Examining the Image

Once we have an image formed by an objective, what do we do with it? We've already seen that the image of a star can be focused on the slit of a spec-

troscope. Professional astronomers often use an image that way, and I've already mentioned that you can use a small spectroscope to see a star's spectrum.

If you place a piece of photographic film or an electronic imaging device exactly at the focus of your objective, you've created an astronomical camera. I'll go into that briefly in Section 4.7.

You can look at the image your objective forms directly with your eye. The problem is that you can't get your eye very close to the image and keep it sharp. There is an optical device that allows you to move your eye close to a small object so that it looks big—the common magnifying glass!

4.3.2.1   The Eyepiece as a Highfalutin Magnifying Glass

The eyepieces in telescopes are special magnifying glasses that allow you to move your eye close enough to the image formed by the objective to examine it very closely. To see what happens with the eyepiece, look again at our simple telescope in Figure 8.

The objective caused the parallel light from the stars to converge at a point in the image plane. The light, being light, can't just stop there and wait for you to examine it. No, it immediately diverges, and would keep on diverging to (what else?) infinity if you didn't put a lens just behind the image.

The eyepiece lens is placed so its focal plane coincides with the image plane of the objective. And the eyepiece lens has a shorter focal length than the objective. The result of the placement of the eyepiece lens (in just the right position) is that the diverging beams of light are made parallel again! Since they are, your eye can look at them as if they were coming at you from an infinite distance. (When a diverging or converging beam of light is made parallel, it is said to have been *collimated*.)

If you look just below the eyepiece lens, you'll notice the "bundles" of parallel light have intersected at one spot. If you place your eye just there, it can take in all of the bundles. This bundle crossover place is called the *exit pupil*, because it looks like the pupil of your eye when you look into the eyepiece of a telescope from some distance away. The distance between the eye lens and the exit pupil is called the *eye relief*. If you place your eye too close to the eyepiece lens, you can still see the central bundle, but then the peripheral bundles can't crowd into your eye. If you place your eye too far from the eyepiece lens, you'll miss the exit pupil, and limit your field of view as well.

The crucial aspect of what the bundles of parallel light are doing is the angle they make with each other. Notice that when the parallel bundles entered the objective, the angles they made with each other were quite shallow. But, when they emerge from the eyepiece lens, the angles are steeper. Since now they look angularly far from each other to the person viewing them through the telescope, the three stars have been magnified.

The magnification of a telescope is the ratio of the angle the stars make with each other when viewed through the telescope to the angle they make with each other when viewed without the telescope. An easier way to determine what magnification a telescope is producing is to divide the focal length of the objective by the focal length of the eyepiece. More on magnification later.

Another way of looking at the exit pupil is that it is an image of the objective lens. When we think about it that way, it makes sense to refer to the objective as the entrance pupil of the telescope. You can prove this to yourself by looking at the exit pupil of a small telescope from a foot or more away, while putting your finger just outside the surface of the objective. You should see a tiny, in-focus image of your finger.

There's another way to determine the magnification of a telescope. Its magnification is also the ratio of the telescope's aperture (the diameter of its objective or entrance pupil) to the diameter of its exit pupil. So, if you have a telescope with an aperture of 100 mm and you measure the exit pupil at 10 mm in diameter, the telescope will magnify 100/10 or 10 times. This is usually written 10×.

4.3.2.2 Matching the Telescope's Exit Pupil to Your Eye's Entrance Pupil

We've already learned quite a bit about the exit pupil of the telescope. We know that you place your eye at the exit pupil in order to ensure that all the bundles of parallel light provided by the telescope's eyepiece can crowd into your eye. There are conditions under which you can't get all the light coming out of the telescope eyepiece into your eye, no matter how well you position it.

One situation is illustrated by the example I gave in the previous section. We had a 100-mm aperture telescope with an exit pupil of 10 mm, giving 10×. If you looked into that telescope, you would see the object at which it was pointed appears to be ten times bigger than it would when you looked at it

without the telescope. However, even when your eye has been perfectly adapted to the dark, it can only open to a maximum aperture of 7 mm. (We say that the iris of your eye is its entrance pupil.) So, when you place your eye at the exit pupil of this telescope, your eye can only accept about 70 percent of the diameter of the exit pupil of the telescope. The rest is wasted.

When you compare the areas of two circles, such as the exit pupil of the telescope and your eye's entrance pupil, simply divide one into the other and square the result. In this case, the answer is $(10/7)^2$ or 2.04. So, even though your eye is accepting 70 percent of the diameter of the exit pupil of the telescope, almost exactly half of the light the telescope is intercepting is not getting into your eye. What can you do?

You can select a power that will result in the exit pupil being 7 mm. Divide 100 mm, the aperture of the telescope, by 7 mm, the desired exit pupil, to get 14.3×. This power will provide the "perfect" exit pupil. (If you can't get an eyepiece that will provide exactly this power, don't worry, just use an eyepiece that provides the next highest power.)

It's no sin to use a power that's low enough to provide an exit pupil bigger than 7 mm. You might do that to see a bigger area of the sky than a perfect eyepiece would provide. However, the eyepiece that provides an exit pupil best matching your dark-adapted eye's entrance pupil will provide the brightest image your telescope can provide.

A note about the perfect 7-mm exit pupil: Children often have entrance pupils greater than 8 mm. Most adults have an entrance pupil of 7 mm. Adults over 50 will notice their entrance pupil gradually shrink, so using a power with their telescopes that produces an exit pupil of 7 mm may no longer be appropriate. Many amateur astronomers of that age use low-power eyepieces that produce an exit pupil of 5.5–6 mm.

The other situation in which you may be unable to get your eye's entrance pupil to coincide with a telescope's exit pupil would be caused by the eye relief being insufficient to allow for the presence of your eyeglasses. There are several solutions to that problem.

1. If you're either nearsighted or farsighted, just look through the telescope without your glasses. Focusing the telescope will make your vision problem irrelevant.

2. Another solution is to wear contact lenses when viewing through your telescope.

3. If your vision problems cannot be corrected by contact lenses, you should know that several manufacturers are responding to the aging of the telescope-buying public by making special binoculars and telescope eyepieces that feature extra-long eye relief. These designs feature an eye relief of nearly 30 mm, which allows room for all but the thickest eyeglasses. Since this feature is a big selling point, you should have little trouble identifying those products.

4.3.2.3 More About Eyepieces

Once you have an objective of high quality, nothing you can do will enhance the performance of your telescope like having a good set of eyepieces. What, you ask, would a good set of eyepieces do for me? This section answers that question.

4.3.2.3.1 About Eyepieces in General

The ideal low-power view through an eyepiece would show you nothing but needle-sharp star images, right to the edge of the field. Only a duck with a bin full of money could afford the optics that would produce such an image. How would we see an image of real stars through real-world optics?

The images in the center of our real-world view would be needle sharp, and they would stay that way halfway out to the edge of the field. They would then gradually fatten-up to the 80 percent line, or 80 percent toward the edge of the field. At the extreme edge of the field, the star images would be fairly large. The dominant aberration here is astigmatism. With few exceptions this is an accurate description of the view through any high-quality eyepiece.

In the next section, I'll go over the pros and cons of eyepiece designs. Before I do, here are some general considerations to keep in mind when selecting a set of eyepieces for your telescope:

1. Angular Field of Eyepieces—When you look into the eyepiece of a telescope you see a field that is (apparently) a certain number of degrees wide. Of all the eyepieces that are widely available, amateur astronomers refer to those that show a 40–50 degree apparent field as *normal-field* eyepieces, while those that show a field of 60–70 degrees are called *wide-angle* or *wide-field* eyepieces. Several manufacturers are marketing eyepieces with a field of about 80 degrees. These are usually termed "ultra-wide-angle" eyepieces.

The real field (or the angular diameter of the portion of the sky you can see) is the apparent field divided by the magnification. If you use an eyepiece with a 65-degree apparent field in a telescope with which it produces 32×, you will see a real field on the sky that is a bit over 2 degrees wide.

Everything else being equal, the wider the field you wish to see, the more money you will spend to see that field. That's because the more field you wish to see, the more you must bend light to do it. The more you must bend light, the more lenses you must have to do it. The more lenses you need, the more money they will cost. These days, some premium-quality, ultra-wide-angle eyepieces can cost as much as a telephoto lens for an expensive 35-mm camera. That's because they are just as complex and are finished as well as a telephoto lens.

The key point about the field of eyepieces is that they will all look like the description I provided above. (Sharp in the center half of the field, with star images flaring out from there.) Essentially, the characteristics of the view are stretched over a wider field. The percentage of field that is perfectly sharp will not change if you view it through a wide-angle eyepiece, but the actual field of sharp images will be larger.

The premium-quality, ultra-wide-angle eyepieces I mentioned above are an exception to this rule. They self-correct their own astigmatism, and their star images are sharp to the edge of the field. Some designs, however, actually give away some sharpness at the center of the field in order to be relatively sharp over the entire field. For planetary observing where a wide field is not important but sharpness is, these eyepieces would *not* be your first choice.

2. Type of Objective—The focal length of a telescope divided by its aperture is called the *focal ratio*. (Example: An objective whose focal length is 900 mm and whose aperture is 60 mm would be known as an f/15 objective.) As you might expect, the smaller the focal ratio, the more light must be bent to make the telescope work. That leads us back to the rule I casually stated in the discussion above about wide-angle eyepieces. I'll restate it in its more complete form. [I (modestly) refer to it as Shaffer's Law of Rabbit-Like Proliferation of Optical Elements.] Here it is:

The more you must bend light in an optical instrument, the more lenses (or mirrors) you'll need to do it, and the more money you'll spend.

As with any law, this one has (apparent) exceptions. However, they usually involve the use of exotic (expensive to buy and fabricate) optical glass, or putting exotic-shaped curves on glass.

Telescopes that feature short focal ratios will generally be much more expensive than those with more moderate focal ratios. For achromatic refractors, a focal ratio of f/15 is usually chosen, but the apochromats can be produced with ratios as small as f/6. Reflectors of f/4 are not uncommon, although producing a high-quality f/4 mirror is NOT a trivial exercise.

However, if you have a short-focal-ratio objective that produces a sharp image, you cannot assume that any eyepiece will work well with the steep angles at which the rays of light diverge from that image. If you have a telescope with a short focal ratio, anything f/6 or shorter, you must ensure that the eyepiece you're considering is designed to handle those steep angles.

4.3.2.3.2 About Specific Eyepiece Designs

You must accept limitations on some aspect of the performance of your telescope in order to minimize the cost of your eyepieces. In general, if you want a wide field and sharp images over much of that field, be prepared to spend a lot to get it. If you want an ultra-wide field and sharp images over all of it, consider investing in a wheelbarrow to carry the money you'll need to the eyepiece store when you make your purchase of a set of eyepieces.

The best way to discuss specific designs is to class them by the number of their elements, or separate pieces of glass, each eyepiece contains.

Two-element eyepieces—There are only two in common use, the *Huygenian* and the *Ramsden*. Both provide narrow fields of view (30–40 degrees), and both suffer from several aberrations. Neither eyepiece should be used with focal ratios smaller than f/10. The Huygenian is often the low-power eyepiece supplied with inexpensive Japanese refractor telescopes.

Three-element eyepieces—There are several, all of which provide a moderate field of 40–50 degrees. The *Kellner* and *Achromatic Ramsden* eyepieces are useful for focal ratios as small as f/7, while some three-element variants of the *Erfle* eyepiece can be used with focal ratios as short as f/4. Many amateur astronomers I know regard a high-quality three-element eyepiece as the minimum acceptable eyepiece for serious viewing through an astronomical telescope.

Four-element eyepieces—There are several types in this class. They are all similar and go by the names *Koenig*, *Abbe*, *Orthoscopic*, *Ploessl*, and *Symmetrical*. The Koenig can actually provide a wide, well-corrected field of more than 60 degrees, but performs well only at focal ratios of f/7 or longer. The others

provide a well-corrected field of 45–50 degrees, and can work well at focal ratios as short as f/4. Many planetary observers will use nothing but high-quality versions of the Ploessl or Orthoscopic eyepieces. Others find that a full set of these eyepieces leaves nothing, except a wide field, to be desired. (Some manufacturers offer a five-element version of the Ploessl, with a well-corrected field of 55 degrees.)

Six-element eyepieces—Most eyepieces containing this number of elements are of the wide-angle type, with very-well-corrected fields of 60–70 degrees. The most popular were designed to work well at f/4. However, there are still military surplus Erfle eyepieces available and some imported Erfle models that work poorly at short focal ratios. These eyepieces are fine for focal ratios of f/7 and longer, though.

Seven–eight-element eyepieces—These eyepieces are fundamentally different from those with fewer elements because they're designed so two or more lens elements are inserted between the objective and the point where the image would normally form. Thus, part of the eyepiece effectively becomes part of the objective. These eyepieces can have apparent fields of about 80 degrees, and the star images will be sharp over virtually the entire field. They were designed for use with almost any focal ratio, and leave almost nothing to be desired.

4.3.2.4 Magnification, Whose Virtues Are Quite Magnified

Earlier, I mentioned that the primary function of a telescope objective is to intercept as much light as possible, concentrating that light into as sharp an image as possible. Then I told you about those special magnifying glasses, eyepieces. All those words about eyepieces might lead you to believe that magnification is the single most important aspect of a telescope's performance, and the more power the better. Aiding that impression might be the ads you see for 60-mm aperture telescopes that are capable of producing 454 power or even (gasp!) 600 power.

Magnification is important but overrated. How well you will see celestial objects is governed by many factors:

1. How sharp the images are that are formed by your objective. If your objective was optimized for a wide field, using it at high power will disappoint you, because the star images will not be sharp.

2. How bright the object is that you're viewing. The more you magnify an image, the more its light

spreads. In fact, if you double its magnification, the light must be spread over an area four times as big, so the image will only be a fourth as bright. Double the magnification again, and the image will be a sixteenth as bright, and so on. A telescope of 60-mm aperture used at 600× will spread the light of an object so much your eye could not detect it! Only the Moon would be even detectable. Although I have a telescope of 20-inch (508-mm) aperture, the highest power I use is 423×, and then only if conditions are excellent.

3. How wide a field you wish to view. It's a fact of physics that, the higher the power you use, the narrower the field. That 600× view through that 60-mm telescope may only show you a field of 0.08 degree, or less than a sixth of the diameter of the Moon.

4. How steady the atmosphere is, a subject covered in the next chapter. I mention it here because it's a factor in how high a power you can use with your telescope.

It's time to give you a strategy for selecting the proper magnification for viewing objects through your telescope— USE THE LOWEST POWER THAT SHOWS YOU EVERYTHING THERE IS TO SEE.

Always view each object first through the eyepiece that gives your lowest power. Look around a bit and center the object in the field. Remove the lowest-power eyepiece and insert the next power up. Is the view an improvement? Do you see more detail? If so, look at it for a while, then boost the power another notch. If not, fall back to the next lower power and concentrate on that view.

4.3.2.5 Selecting a Range of Eyepieces

In this section, I'm not going to recommend a specific type of eyepiece for your telescope, but I am going to recommend a specific range of powers. This range is based on three considerations:

1. The aperture of the adult dark-adapted eye— 7 mm. We'll use this as the lowest power. As we've seen, using a lower power only results in less of the light intercepted by your telescope being admitted into your eye.

2. The human eye seems to resolve fine detail best when operating at an aperture between 0.5 and 1.0 mm. Even if our eyes are opened up to 7 mm, if we use a power that results in an exit pupil of

1 mm, we use only the central 1 mm of our eye. So, we are effectively limiting the aperture of our eye to 1 mm. We'll select as our highest powers ones that result in exit pupils in the range of 0.5–1.0 mm.

3. Remember, when you double the power of any telescope, you cut the intensity of the image to a fourth of what it was. That is too much of a jump. An ideal range would be spaced so that each time you raise the power, you cut the light intensity in half. That means each power will be 1.4 times the previous.

Here's the range of powers for telescopes of 60-mm to 508-mm aperture (2.4–20 in.):

Telescope Range of Powers										
Exit Pupil (mm):	Aperture (mm):									
	60	80	100	115	152	203	254	320	406	508
7	9X	11X	14X	16X	22X	29X	36X	46X	58X	73X
5	12	16	20	23	30	41	51	64	81	101
3.5	18	22	28	32	44	58	73	91	116	145
2.5	24	32	40	46	61	81	101	128	162	203
1.75	36	44	56	64	88	112	144	184	228	290
1.25	48	64	80	92	122	162	202	256	324	406
0.88	72	88	112	128	176	224	288	368	456	2Hi
0.63	96	128	160	184	244	324	404	512	2Hi	2Hi
0.44	144	176	224	256	352	448	2Hi	2Hi	2Hi	2Hi

When obtaining a set of eyepieces for your example telescope, you might decide how much you can spend, then select the highest-quality type of eyepieces that will result in your obtaining as many of the powers on the above list as possible.

Or you might consider delaying the purchase of the whole set and select a smaller range to start. That might involve not selecting the lowest power eyepiece, then purchasing every other power, filling in the others later. Another strategy might be to select a larger multiplier than 1.4, covering the range between 7 mm and 0.5 mm exit pupils with fewer eyepieces. If you are over fifty, you might start the range at 5.5 or 6 mm, since it is likely your dark-adapted eyes will not quite open to 7 mm anymore.

NOTE: There's no rule that you must exactly mimic this table. In fact, it is unlikely that you will find any set of eyepieces that would provide such an exact progression of powers. However, the closer you can come to that progression, the likelier you will be able to view all objects at nearly the best power at which they can be viewed.

Finally, you might say to yourself, "Why not use a zoom eyepiece, and stop fiddling with selecting and changing an eyepiece at all? Just dial in the appropriate power and be done with it." Good question. The answer is twofold: The zoom eyepieces on the market today have narrow fields and cover an insufficient range of powers. It also seems that few potential customers are willing to pay the large initial price for a premium-quality eyepiece that would cover the above range of powers.

4.3.2.6 Helping the Eyepiece Do Its Work: The Barlow Lens

So far, we've only considered lenses and mirrors that focus light. Another way of looking at a lens that focuses light is that it takes a parallel beam of light and causes it to converge to a point. Converging lenses are "fatter" in the middle than at the edge, and converging mirrors are concave.

Any lens that diverges light has several uses; among the first was in telescopes as the eyepiece. If a diverging lens is placed at the right position inside the focus of an objective, it makes the converging beam coming from the objective parallel again. If you place your eye just behind the lens, you will see a magnified view that is right side up! That was the original arrangement of lenses in Galileo's telescope. Although it is simple and features an erect image, you can't see a very wide field with a Galilean telescope, and you can't produce a very high-power telescope with it, either.

Instead of using a diverging lens to recollimate the beam coming from an objective, a British astronomer named Barlow used such a lens to stretch the beam into a much shallower cone than it would have been without his lens. The resultant image is quite a bit larger than it would be without the Barlow lens. Since the power of a telescope is determined by the size of the image, the Barlow lens is really a telescope power booster.

Some Barlow lenses can boost power anywhere from 1.8 to 3 times. They are usually supplied in a tube that is inserted into the focuser of the telescope. The end that sticks out of the focuser is machined to accept the telescope's eyepieces.

You may already have realized that selecting a Barlow lens of the right magnification will allow you to use each of your eyepieces to obtain two powers with your telescope. If you check the table in Section 4.3.2.5, you'll see you could equip your telescope with eyepieces that provide the first three powers in the table, then use a Barlow lens that magnifies

2.8 times to obtain the next three powers. Since a good Barlow lens will cost as much as the same-quality eyepiece, you would be obtaining six powers for the price of four!

But wait, there's more. The Barlow lens also makes it easier for your eyepieces to form an image. That's because the Barlow has made a steep cone of light into a shallow one. Since the cone is shallow, the eyepiece doesn't have to bend the light through so large an angle and will do the job better. The result: a sharper image.

There's one final advantage to the Barlow lens. It increases the eye relief on any eyepiece with which you use it. If you wear glasses, you may want to get friendly with a good Barlow lens.

(NOTE: The seven–eight-element eyepieces I described in Section 4.3.2.3.2 use a group of diverging lenses as a built-in Barlow lens and to correct some aberrations of the objective. These eyepieces also tend to have a longer eye relief than others of the same focal length.)

4.3.2.7 About Eyepiece Plumbing

Telescopes usually have a device for moving the eyepiece in and out along the optical axis so you can focus the telescope. It can be as simple as a tube through which the eyepiece slides, or as complex as motorized devices costing as much as 50-yard-line seats at the Super Bowl. Most focusers are designed to accept one of three standard sizes of telescope eyepieces.

Smaller telescopes often come equipped with eyepieces whose barrel is 24.5 mm, or 0.965 in. in diameter. Because telescopes in this class seldom have large apertures, the selection of eyepieces in this barrel diameter is limited. Lately, however, the larger dealers in telescopes from the Orient have begun importing beginner telescopes of higher quality than before. These are equipped with 0.965- in. eyepieces. These same importers are also offering an expanded selection of high-quality replacement eyepieces.

For years, the American standard eyepiece came in a barrel diameter of 1.25 in. The size was chosen in the fifties, because amateur astronomers didn't like the narrow field of view offered by the early microscope eyepieces then in use. An extremely wide range of eyepieces and other accessories can be found in the 1.25-in. eyepiece barrel diameter.

As the sixties dawned, eyepieces with a barrel diameter of 2.0 in. were introduced. This size actually came about because amateurs wanted to use lower and lower powers to examine very faint objects and to obtain a wide field while doing so. There is a wide range of eyepieces and accessories available in 2.0-in. sizes.

4.3.3
What the Wave Nature of Light Does to an Image

The most obvious consequence of the wave nature of light can be seen when we point a telescope at a bright star, then examine the star image at moderately high power.

We know the stars are so far away that their angular size is, essentially, zero. If the telescope is high quality and the atmosphere is steady, we might expect the star's image to be an incredibly tiny point of light. Instead, we see a tiny disc of light, surrounded by two or three rings of light, each of which is fainter than the next, until the farthest one out is nearly indistinguishable. If we increase magnification, the disc and rings become bigger, thus dimmer, but we see no further detail.

Is what we see an image of the star? No, it's the diffraction pattern produced by the telescope, which is caused by the wave nature of light. This pattern is caused by light waves passing by the edge of the lens or the mirror in your telescope. They are said to be diffracted by the edge of the objective. The disc at the center of the pattern is called the *Airy disc*, after British astronomer Sir George Biddell Airy, the same man who invented the projection used for the ALL-Sky maps. You can read further on this subject in a book on telescope making.

The fact that we see this diffraction pattern rather than a perfect image of a star has a profound effect on the performance of a telescope. It affects a telescope's performance in two principal ways.

4.3.3.1 Diffraction Limits Resolution

Suppose you point your telescope at a double star, employing a very high power. What will you see? If the separation of the stars is large, you'll see their separate diffraction discs, surrounded by a jumble of fainter diffraction rings. As you look at stars separated by smaller and smaller angles, those patterns begin to merge. When they are superimposed so that the edge of one diffraction disc coincides with the center of the other, you have reached the resolution limit of your telescope. You see the two discs as a "notched-double," surrounded by their almost-merged diffraction rings.

A telescope that can resolve this phenomenon well is said to satisfy the *Raleigh Criterion*. The equation that provides the separation of two stars that will just meet the Raleigh Criterion is:

Resolution = 140/(aperture in mm)
OR 5.5/(aperture in inches)

Regardless whether you use millimeters or inches, the answer is in seconds of arc. (An astronomer named Dawes determined experimentally that experienced observers could distinguish stars separated by an angle about 20 percent smaller than that given by the Raleigh Criterion. If you put in a lot of time training your eye-brain combination to discern small differences in light level, you might graduate to using the *Dawes Limit* instead of the Raleigh Criterion.)

Philosophical aside: Before you curse whoever decided to include diffraction in the "Design of the Universe," consider this: If diffraction did not spread the image of a star into the diffraction disc, a star image would be an incredibly small disc. In other words, we would have a true image of that star. That image would be as bright, and as hot, as the actual surface of the star! Unless we had evolved differently, we wouldn't be able to look at the starry sky without injuring our eyes.

4.3.3.2 Diffraction Reduces Contrast

If we take a closer look at the diffraction pattern, we see that most of the light the telescope has intercepted from the star is concentrated in the disc. In fact, a perfect objective places 84 percent in the disc, 7 percent in the first ring, 3 percent in the second, and the rest in the third, fourth, fifth, and so on. The rings go on indefinitely, but there is so little light in those past the second that, fortunately, we can't see them.

If we view an extended object such as Jupiter through a telescope, each point on the planet causes a diffraction pattern to be present in the image. Thus, the image we see is really caused by all those individual diffraction patterns superimposed on one another. Those features that are very close together are not resolved, because their diffraction patterns are on top of each other. We can't tell the two patterns apart.

However, the fact that all those individual sets of rings are superimposed on one another also affects what we see, because they make it harder to resolve features whose angular size is small. They do that by limiting the contrast.

Besides being a knob on your TV, what's contrast?

Suppose there were a pair of straight black lines on a perfectly white planet. If we were hovering right above them, we would see them exactly as they are. Since the alternating lines are either perfectly white or perfectly black, we say that the contrast between them is 100 percent. A more universal definition of contrast is to say it is the "ratio of the brightest to the dimmest."

Now let's view our pair of lines from Earth through a telescope. We would see the lines, but their edges would be gray, since the diffraction rings of each point in the white areas next to the black lines are superimposed on the line itself. So, as we see it, some of the white has "leaked" into the black. Diffraction has reduced the contrast between the lines and their surrounding white areas and they don't look as sharp as we know they must be.

Now let's gradually move our lines closer and closer together. We find that, even though the object contrast remains at 100 percent, the image contrast we see drops more and more, as the lines get closer together. At some point, the contrast drops below the ability of our eye-brain combination to detect and the pair of lines becomes invisible to us.

For objects with 100 percent contrast, the lower resolution limit imposed by the reduction in contrast I've just described is not much different from the Raleigh Criterion. However, if the object contrast is less than 100 percent, as you might find in the subtle details of a planet's features, the resolution limit imposed by a reduction in contrast is considerably greater than the Raleigh Criterion.

Therefore, a telescope cannot resolve low-contrast detail on the surface of a planet nearly so well as it can resolve double stars. If you want to use your telescope to observe planetary detail, it's important to make sure that nothing happens to reduce contrast.

What reduces contrast?

1. Dirty optics scatter light all over the place. This is no different from what happens to you when you're driving toward the setting sun and your windshield needs cleaning.

2. An objective lens or mirror whose shape deviates from perfection causes light to find its way from the diffraction disc into the rings, thus reducing contrast.

3. Anything that obstructs the light path to the objective causes more diffraction, which allows more light to be distributed into the rings. That reduces contrast. I'll go into this later in the discussion of reflector telescopes.

4. An unsteady atmosphere does many things to your view, including reducing the contrast. More on that in Chapter 5.

4.3.4
Types of Telescopes

The reason I haven't discussed the various types of telescopes in greater detail until now is that I thought you needed to understand first how a telescope—any telescope—functions.

Whole books have been written on this subject, so I'll cover the high spots.

NOTE: To simplify the diagrams in Figure 9, I've drawn only a few of the rays of light from a star on the axis of each example and I've omitted rays of light from stars that are off-axis.

4.3.4.1   The Refractor

The refractor was the first telescope. It probably resulted when someone was playing with various spectacle lenses and noticed that a magnified view of a distant object could be seen when two different lenses were held in the proper relative positions. The first telescopes were possibly made in Holland by spectacle makers sometime in 1608. By 1609, Galileo, an Italian scholar, had heard of the telescope and made one. He is the first person to have systematically observed the sky with a telescope, and he made discoveries that, literally, "turned the world on its ear."

Galileo observed that small objects, Jupiter's satellites, visible only through his telescope, were obviously in orbit about a big, bright object, Jupiter itself. That evidence refuted the theory that the Earth is the center of the Universe. Since that theory was part of basic religious teaching, Galileo found himself at odds with the Catholic church. You can read more about this feud in astro-history books.

What's more intriguing about the observations Galileo made nearly four hundred years ago with his telescope is that he might well have killed to have the worst telescope available in most department stores today. And other scholars might have helped. Despite his telescope's limits, Galileo discovered that the Moon has craters and mountains on its surface, that the Sun is blemished by spots, and that Jupiter has bands around its disc and four satellites orbiting it. Another astronomer, Cassini, discovered the rings of Saturn.

Figure 9: Types of Common Telescopes

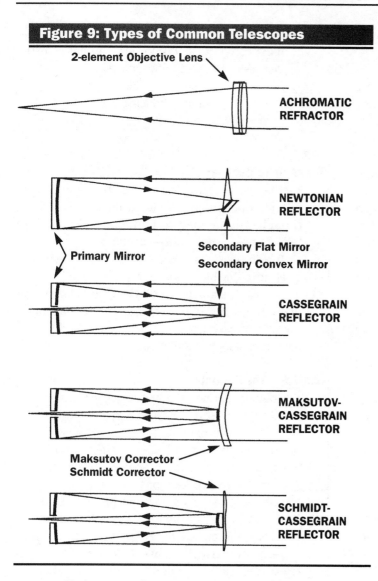

All this was done with a telescope whose objective lens was only a bit more than an inch in diameter that magnified only about thirty times. Generally, the smallest astronomical refractors available today have apertures of 50 or 60 mm, twice Galileo's, and their lenses are achromatic so the images are almost perfectly corrected for stray color. Finally, they come with eyepieces that give a modest range of powers, rather than the single power Galileo had to use.

Because the aperture of a refractor is unobstructed, the contrast of the images it produces is as high as it can be. Refractors provide sharp, clear images of the Sun, Moon, and planets, and are generally favored by observers who specialize in observing those bodies. However, since a refractor must have a minimum of two lens elements in its objective, it is much more expensive to manufacture than the simplest kind of reflector telescope. Also, a refractor must be physically

longer than a reflector, since having a long focal length is one element in minimizing chromatic aberration. (The exception is the apochromatic refractor, which can have a focal ratio as short as f/6.) Because of the size and complexity of refractors, they are seldom seen in sizes larger than 7-in. apertures, except in large observatories operated by schools and planetariums.

4.3.4.2   The Newtonian Reflector

While the refractor was certainly discovered by accident, the reflector telescope was designed theoretically before the first one was built. In a burst of activity from 1665 to 1685, three types of reflector telescopes were designed by several mathematicians. The simplest is known as the Newtonian reflector after its inventor, Sir Isaac Newton, the genius who also discovered the law of gravity.

Newton was trying to design a telescope that would not suffer from chromatic aberration, as the achromatic refractor telescope was yet to be invented. His telescope features a concave mirror in the shape of a paraboloid, located at the bottom of a tube. Light from a star enters the tube, strikes the mirror, and is brought to a focus near the open end of the tube. To avoid the optical path being completely obstructed by the eyepiece and the head of the observer, a flat mirror or prism is placed just inside the focus. The diagonal prism or mirror places the focus just outside the tube, where it can be examined with an eyepiece.

Newton's telescope was well received, but, until another scientist, Foucault, invented a test that allowed the subtle figure of the paraboloid to be placed on the mirror's surface, reflectors were difficult to make. Even today, figuring a paraboloidal mirror is considered a nontrivial exercise. Also, the shorter the focal ratio of the paraboloid, the more difficult it is to figure.

Theoretically, the on-axis image of a star provided by a Newtonian reflector is perfect. Since light of all colors is reflected equally, a reflector suffers no chromatic aberration. Only diffraction limits the perfection of the on-axis image of a star seen through a Newtonian reflector.

Once a star's image departs from the center of the field of a Newtonian, we say its rays are now off-axis. The farther the image is from the center, or the farther off-axis its rays are, the more the image deteriorates. The aberration that causes this is coma, one of the off-axis aberrations I discussed when we were going over telescopes generally.

Coma severely limits the field of a Newtonian reflector. Fortunately, the larger the focal ratio of a Newtonian, the smaller its coma. However, the price we pay for this is that the long-focus Newtonian is quite large and shows a very narrow field.

The two main advantages of a Newtonian reflector over a refractor are that the reflector is perfectly achromatic and that there is only one optical surface to figure, rather than the four to six that must be figured with a refractor. In the larger sizes, a reflector really comes into its own as its relatively short tube length becomes more and more important.

Reflectors have often been given a "bad rap" because they often seem to give poor images. It's been my experience that the poor images have two causes:

1. A poorly figured mirror. In other words, the mirror does not have the proper shape (the paraboloid).
2. The telescope's owner has, simply, improperly aligned the optics so the optical axis of the objective mirror is coincident with the optical axis of the eyepiece. (This process is called collimation. You can read a description of it in any book on telescope making.)

These two reasons why reflectors often perform poorly are not inherent to the design. The chief disadvantage of a reflector that is genuinely built into the design is the loss of contrast in the image caused when the diagonal mirror and its supports obstruct the aperture. I mentioned in the previous section that for a perfect, unobstructed objective, 84 percent of the light from a star is found in the Airy disc and 16 percent is distributed in the diffraction rings. For a perfect objective mirror in a reflector with an obstruction 25 percent as wide as the mirror, only 73 percent of the star's light makes it onto the Airy disc, and 27 percent goes into the rings. A reflector with a 33 percent central obstruction will lose another 5 percent to the rings, so that only 68 percent of the light it receives is actually found in the Airy disc.

This loss of contrast means the image of a planet seen in a refractor will seem to be sharper than the same image in a reflector of the same aperture. However, this disadvantage can be overcome by using a slightly larger reflector. If you like the view through a high-quality 6-in. refractor, you may find that the view through an 8- or 10-in. reflector is as

good, at a significant reduction in cost. (This statement is sure to be contested by the large and vocal Refractor Appreciation Society.)

4.3.4.3 The Cassegrain Reflector

The Cassegrain reflector uses two curved mirrors to form an image. The first is the familiar paraboloid used in the Newtonian, but it usually has a short focal ratio. Another mirror is placed in the upper end of the tube just inside the focus of the primary mirror. This mirror is convex and its figure is another nonspherical curve, the hyperboloid.

The Cassegrain secondary mirror performs the same function as a Barlow lens: It takes the steep cone of light coming from the objective mirror, or primary mirror, and stretches it out into a much longer, shallower cone. The final image is formed behind the primary mirror, so the primary must have a hole in it. There also must be a tube just in front of the primary mirror to prevent light passing just next to the secondary from flooding the eyepiece.

The advantages of the Cassegrain are twofold: (1) It has the capability to obtain high powers in a telescope that is quite compact, and (2) the position of the eyepiece is much more convenient than on the Newtonian.

But there are some important disadvantages to the Cassegrain design:

1. It's very difficult to figure the two mirrors, both of which are complex, nonspherical curves. This consideration caused the father of amateur telescope making, Russell W. Porter, to entitle a chapter in *Amateur Telescope Making*: "How to Make a Cassegrain Telescope (and Why Not to)." (The two-mirror arrangement used in the Hubble Space Telescope is a variation on the basic Cassegrain design.)
2. The collimation of the two mirrors in a Cassegrain telescope is critical. A Cassegrain will often perform badly due to misalignment of its optics.
3. A Cassegrain telescope cannot be made to have both a wide field and one in which no stray light is present. This is because the baffle that eliminates the stray light also limits the field.
4. Since it must have a central obstruction, a Cassegrain suffers from the same loss of contrast as the Newtonian. However, if a very narrow field of view is acceptable to you, the central obstruction can be kept modest, and the loss of contrast would be moderate.

4.3.4.4 Hybrids of the Refractor and Reflector

There are a wide range of telescope designs making use of a combination of lenses and mirrors to form an image. The word for such designs is "catadioptric," although most people shorten that to "cats." They are often called "mirror-lenses" by photographers. There are several variations, but there's one important point: The arrangement of the mirrors in "cats" is the same as if the lenses were not part of the design. With few exceptions, these hybrids are variants of either the basic Newtonian or Cassegrain designs.

The lenses in catadioptric reflectors are included in the design because they allow the correction of an aberration that couldn't be corrected without them. These lenses are usually called "correctors," and fall into two classes: (1) full-aperture correctors that intercept the light from a star before it reaches the primary mirror, and (2) sub-aperture correctors that modify the converging cone of light after it has struck the primary.

Most of the "cats" available on the amateur market are full-aperture types. However, several accessory sub-aperture corrector lenses are available either to reduce or remove the coma that is the chief aberration from which a Newtonian suffers. Also, the seven- to eight-element eyepieces I described in Section 4.3.2.3.2 could be considered cousins to the catadioptric, because they employ several lenses to correct aberrations of the objective before the other lenses in the eyepiece magnify the image.

There are "Catadioptric-Newtonians" available, too, but they make up a tiny portion of the telescopes used by amateur astronomers. There are, however, MANY examples of the "Catadioptric-Cassegrains."

4.3.4.4.1 The Schmidt-Cassegrain Telescope

Both the Schmidt-Cass and the Mak-Cass, as the two predominant "cats" are known, are developments of astronomers' attempts to make reflector telescopes that do not suffer from coma. The first of these types of telescopes was really a camera, developed by an eccentric genius named Bernhard Schmidt in the thirties. (Schmidt lost an arm to youthful experimentation with rocketry. Fortunately, his efforts in optics were more successful.)

Schmidt's corrector is placed at the sky end of the telescope's tube and is almost flat. The corrector plate has a very shallow curve often called a "dough-nut." (In Figure 9, I've exaggerated the doughnut

hundreds of times so you can see it.) Placement of the corrector plate determines how deep the doughnut curve must be. In the Schmidt camera, the corrector is placed at the center of the sphere defining the surface of the reflector. The film is placed halfway between the center of curvature and the mirror. To achieve sharp focus, the film must be warped into a spherical surface itself. That may seem a lot of bother, but it's worth it. Schmidt cameras provide exquisite wide-field photographs of the sky and there's no coma!

The trick to the Schmidt corrector is that it has a lot of spherical aberration, but if figured just right, its spherical aberration is equal and opposite to the spherical aberration the mirror would produce. They cancel.

The Schmidt-Cassegrain is a development of the Schmidt camera. The Cassegrain mirrors can be made spherical, rather than the more complex curves that the conventional Cassegrain requires. Since the mirrors are spherical, there is no coma. There is, however, a lot of spherical aberration. As you can guess, the corrector used in the Schmidt-Cass has just the right doughnut curve to cancel it.

Today's Schmidt-Cassegrain telescope is a compromise between perfect optical correction and compactness. To shrink the length of the tube, the corrector is not located at the center of curvature of the primary mirror, but just in front of the secondary mirror (Figure 9). The Cassegrain secondary is mounted on the Schmidt corrector. Since the corrector is closer to the primary than it would be in an optimum design, the secondary mirror is usually made slightly ellipsoidal. (An ellipsoid is a surface that's somewhere between the sphere and the paraboloid.) This slight change counteracts the spherical aberration caused by moving the corrector inward.

Since it became widely available in the early seventies, the Schmidt-Cassegrain telescope, or SCT, has become the workhorse of the amateur astronomer. It is made in 4-, 5-, 6-, 8-, 9-, 10-, 11-, 12-, and 14-in. apertures, usually with a focal ratio of f/10 or f/11. Since the focal ratio of the primary mirror in all of these designs is usually f/2, the length of the SCT's tube is only twice the mirror diameter, yet its focal length is that of a more conventional telescope whose tube is ten times as long as its mirror diameter.

There is an amazing variety of brands and sizes of SCTs on the market, with the 8-in. model leading in popularity. In fact, it is likely that there are more SCTs in the hands of amateur astronomers in North

America than all other types of telescopes combined. There is an equally vast selection of accessories available for the SCT.

Almost all of the available SCTs use a different method of focusing than I described in Section 4.3.2.7. Instead of moving the eyepiece in and out to achieve focus, the primary mirror of the typical SCT is moved in and out along the baffle tube. This allows the image to be moved relative to the back plate of the SCT. The range of focus is something in the neighborhood of 8 inches, which allows the owner of an SCT to focus almost any eyepiece or camera she or he can find. That fact alone allows the amateur to add whatever "plumbing" is necessary to do whatever she or he wishes.

The SCT is an incredibly versatile instrument. The amateur can view planets with it. It does reasonably well at viewing the dimmer deep-sky objects, and it does well at photographing them. It is the telescope for all seasons. Because the tube of an SCT is so short, it is light and doesn't require a massive mount to allow it to point to any position in the sky. A complete SCT in all but the largest sizes is truly a one-person telescope; any adult can unload it from a small car and set it up without assistance.

When I am asked to recommend an all-purpose telescope of relatively large aperture, I almost always recommend the 8-in. SCT. It provides beginning amateur astronomers the most versatile, compact, easy-to-use astronomical telescope.

However, the SCT is not the finest telescope available. There are apochromatic refractors and custom-Newtonian reflectors available that will provide much sharper images than an SCT. Why do I recommend the SCT?

1. VERSATILITY—SCTs can do just about anything an amateur wishes them to.

2. COMPACTNESS—Many amateur astronomers live in urban areas, where one of the many types of pollution is extraneous light in the dark sky. If we truly want to see the sky as we were meant to, we must travel several hours to get out to the boonies. The SCT is the most telescope (for the money) that will both fit in your car and be easy to set up.

3. PRICE—The premier instruments mentioned above all cost far more than the SCT. Also, they are not readily available, as they are not mass produced.

Everything I've written can be boiled down to one word: compromise. Amateur astronomers buy (and should buy) SCTs for much the same reason a person might buy a station wagon or a compact pickup truck as her or his only car: because it does many tasks in a satisfactory manner, even though it does nothing very well. Put another way: If you knew you were going to be hauling lumber home from the lumber yard pretty often, would you buy (as your only car) a Ferrari?

4.3.4.4.2 The Maksutov-Cassegrain Telescope

The Maksutov-Cassegrain telescope is identical to the Schmidt-Cass, except the almost-flat corrector plate is replaced by a deep-dish lens that corrects the spherical aberration of the Cassegrain mirrors. (The corrector lens is called a meniscus lens. Because it is no thicker at the edge than at the center, it neither diverges nor converges the light passing through on its way to the primary mirror. It only corrects spherical aberration.) Often, the secondary mirror is eliminated and a spot on the back of the corrector is aluminized instead.

The Maksutov corrector is more expensive to make (in mass production) than the Schmidt corrector. It is also quite thick and in the larger sizes quite heavy. For these reasons, Maksutov-Cassegrains are considerably more expensive to manufacture than SCTs. "Maks" are not found in nearly the numbers as SCTs. Those that are available are usually of premium quality, delivering images that leave little to be desired.

4.3.4.5 Finder Telescopes and Other Pointing Devices

A finder is a telescope with very low power and thus a very wide field. A finder usually has a glass reticle at the focus of its objective. The reticle is placed there so it can be viewed at the same time as the view through the finder. Etched on that reticle is a pattern, usually cross hairs, that defines the exact center of the finder's field of view. The finder is mounted on the tube so that it can be moved slightly to align its cross hairs on the object that is simultaneously in view through the main telescope. (A telescopic sight on a rifle is nothing more than a specialized finder.)

To align your finder, locate a very distant terrestrial object in the main telescope using low power. Lock the telescope in position, then adjust the finder until the object you see in the main telescope is centered in the cross hairs of the finder. (Normally this is done by loosening the screws holding the finder in the rear support ring, then tightening them gradually to center the cross hairs on the target.)

There's no rule that you must have a finder. Some small telescopes can operate at such low power that they do not need a finder. Other telescopes feature devices known as reflex sights that use optics to "project" a red-lighted bull's-eye pattern onto the sky. You just look up at the sky and move the telescope until the bull's eye is in the right spot. These devices are very popular and may be found on maybe one of four amateur telescopes.

4.3.5
How to Look Through a Telescope

In the previous sections of this chapter, I've already mentioned several factors to take into consideration when looking through a telescope.

If you wear glasses to correct your vision for nearsightedness or farsightedness, take them off. When you first look into the eyepiece, place your eye in the middle of the lens opening. Move your eye in and out until you can see the widest view. Lightly grasp the focus knob and slowly turn it until the sharpest image is obtained. While you focus the telescope, relax your eye, and "tell it" you're looking at an object far away. If you're looking at a dim object, try looking at it out of the corner of your eye so the greater sensitivity of your rods can see it better. Finally, take your time and enjoy the view.

4.3.6
Testing Optical Performance

As you might imagine, whole books can be written on this subject. However, there's one test that will give you a fairly good idea how accurate the optical system of a telescope is. Just follow these simple steps:

1. Point the telescope at a bright star as near the zenith as possible.
2. Use a high-power eyepiece to examine the star's image. If the seeing is really bad, you may have to postpone the test until a night when it's steadier.
3. Slowly change the focus of the telescope so that the star image becomes a disc, and then slowly move the focus of the telescope through the point of best focus. Do this several times.
4. The optical system is at least "O.K." if the images of the star on both sides of the focus are identical. (If the disc on the inside of the focus is brighter at its edge than at its center, the disc at the outside will likely be brighter at its center than at its edge. That's a sure sign of spherical

aberration, and a poorly corrected optical system. If the discs are not perfectly circular, the system has an even worse problem, called astigmatism.)

5. On a night of very good seeing, this test can tell you even more, but only if you have considerable experience. However, even in the hands of a beginner, it can usually spot a telescope that's sub-par. It won't allow you to discriminate between a good and an excellent telescope without the aforementioned experience, however.

4.4
Mounting a Telescope So It Can Move About the Sky

With the finest optics, you still won't see much if the support for your telescope can do an imitation of Santa Claus's belly ("quiver like a bowl full of jelly"). That support is called the mount and it has several purposes. (1) Not to vibrate too much when you touch your telescope to move it or focus it. (2) To let you easily point your telescope to any point in the sky. (3) Either to allow you to smoothly track an object by hand OR automatically to track an object with the aid of a motor.

When selecting a mount, your watchwords must be smooth and sturdy. Those watchwords apply equally to the tripod or other support as well.

4.4.1
The Altazimuth Mount

In Section 2.5.3, I wrote about the coordinate system that we call the local sky. An altazimuth mount moves along that coordinate system. It turns on two axes that allow it to move in one direction along the horizon ("in azimuth") and in the other direction perpendicular to the horizon ("in altitude"). Given the choice of "azimalt" or "altazimuth," astronomers chose the latter.

The altazimuth mount is simple and easy to use. In smaller sizes, it can be found supporting inexpensive refractor telescopes. Often there are slow-motion controls on each axis so these simple altazimuth mounts can be moved a few degrees in any direction from their original position. These controls can also be released so the telescope can be quickly slewed a long distance across the sky.

For larger telescopes, especially huge Newtonians, there is a variant of the altazimuth called the Dobsonian, after the former Vedantan monk, philosopher, and telescope maker, John Dobson, who invented it in the late sixties. The Dobsonian has three features to recommend it.

1. It can be made with commonly available materials by persons who lack machinist skills.

2. It can be moved easily around the sky by hand, even in the largest sizes, yet it can be moved a tiny fraction of a degree with equal ease.

3. Its design essentially eliminates a tripod by placing the azimuth axis almost on the ground. This puts most of the telescope's weight close to the ground. The whole structure is a strong semi-box. Its design, and the fact it is almost always made of wood, means the Dobsonian mount vibrates radically less than other mount designs.

The Dobsonian-mounted telescope has become extremely popular, and the altazimuth mount in general has a lot to recommend it. However, there is one thing no altazimuth mount can do: easily follow the stars in their paths across the sky. To do that, the altazimuth mount must make a continuous herky-jerky, or stair-step, motion in both azimuth and altitude. Between these motions, the object in view seems to drift across the field. And, of course, the higher the power you use, the faster the object drifts across the field.

What to do? There are several rather esoteric solutions. One involves controlling the motion of your altazimuth mount with a computer and small motors. For Dobsonians, there are devices known as equatorial platforms on which the whole telescope is placed, which move the same way the sky does. The simplest solution is often to make an equatorial mount, the subject of the next section.

4.4.2
The Equatorial Mount

We'd like a mount that allows us to easily follow any object on the Celestial Sphere as it rotates. This mount would make any object appear to stand still in the field.

Even a casual reading of Chapter 2 probably persuaded you the Celestial Sphere appears to rotate around the Earth's axis. Since everything on the Celestial Sphere (except the Moon) is incredibly far away, the distance between where we are on the surface of the Earth and the Earth's axis is infinitely small, or, in effect, zero. From looking at the Celestial Sphere, we can't tell much difference between being on the surface of the Earth and being at Earth's axis.

You may already have realized we can make the

mount we seek by having its principal axis of rotation be parallel to the Earth's axis. This axis is the *polar axis*.

Since the polar axis is parallel to the Earth's, and since we can assume we are at the Earth's axis, all we have to do is rotate our polar axis at a rate equal in speed and opposite in direction to the rate at which Earth's axis rotates. (Another way to think about this: In order to track the stars, we can rotate our polar axis just as the Celestial Sphere rotates.)

We say our polar axis moves in right ascension. By rotating our polar axis, we can move our telescope east and west on the Celestial Sphere. (Remember that right ascension on the Celestial Sphere is the same as longitude on Earth.)

To move our telescope north and south (in declination), we must have another axis at right angles to the polar axis and attached to it. When these two axes are arranged this way, we can point our telescope at any point in our sky.

What I've described is called a polar equatorial mount, which most astronomers call an "equatorial" for short. Because the polar axis must be parallel to Earth's axis, it must be tilted with respect to your horizon. In fact, when a polar axis is properly aligned, it points exactly at the North Celestial Pole (for observers in the northern hemisphere).

You can think of an equatorial mount as an altazimuth tilted over to point at the pole. That's quite true, and some simple equatorial mounts are little more than that. However, you should know that because its principal axis is tilted away from the vertical, an equatorial mount is loaded unevenly by gravity. To vibrate as little as a stout altazimuth mount, an equatorial must be made beefier than an equatorial mount. More beef means more money, so expect to pay more for an equatorial mount than for the equivalent altazimuth.

Even if you have no automatic means of slowly rotating the polar axis of your mount, it's simpler to track an object manually than with an altazimuth. However, as I've pointed out, the equatorial is more complicated to make sturdy and vibration-free than the altazimuth. You might very well choose the altazimuth, especially the Dobsonian, if you were not going to automatically drive the polar axis of an equatorial.

4.4.2.1 Types of Equatorial Mounts

There are really four types of equatorial mounts, however, most equatorials are either the German equatorials or fork mount.

4.4.2.1.1 A GEM of a Mount

The German Equatorial Mount is shown in Figure 10. It apparently got its name because it first came into wide use in Germany.

The GEM, as it is sometimes called, features a polar axis on top of the tripod or pier. The declination axis is at the upper end of the polar axis. Attached to one end of the declination axis is the tube of the telescope, at the other end is a weight to counter the weight of the telescope. This counterweight is usually able to be moved up and down a shaft or threaded rod to allow for exact balance of the tube and mount around the polar axis. This balance is critical to avoid overloading the drive motor.

The German Equatorial Mount is compact and if carefully designed and built, will support a telescope with a minimum of vibration. Unless the tube collides with the tripod or pier, the GEM allows the telescope to point to any point in the sky. It is in wide use.

No mount is perfect. The disadvantages of the GEM are that the telescope cannot track an object continuously through the meridian. At some point, the telescope will collide with the tripod or pier, then the mount must be rotated 180 degrees to the other side of the polar axis to permit continued tracking. The counterweight also adds weight to the package. Furthermore, if you've either forgotten your counterweight or dropped it on your foot, you may not remain a fan of the German Equatorial Mount. (In my lifelong search for completeness, I've done both.)

4.4.2.1.2 The Fork Mount

The Fork Mount uses a similar polar axis to the GEM, and its declination axis is also placed on the upper end of the polar axis. However, the declination axis is placed at the end of a fork-shaped assembly, and the tube swings around that axis between the two tines of the fork.

The Fork Mount is generally more convenient to use than the GEM because the telescope can track through the meridian. Also, this type of mount is symmetrical around the polar axis, so the counterweight is unnecessary. (Because it integrates nicely into the overall concept of the Schmidt-Cassegrain telescope, the fork mount is the one chosen for the vast majority of commercially available SCTs.)

Because the fork mount must allow the tube to swing through it without colliding with the throat of the fork, the tines of the fork must be relatively long. If they are not made quite large, they will vibrate

excessively. Also, the fork mount is considerably less compact than a GEM of the same level of performance.

4.4.3
Using a Mount

Properly set up, a good mount will support your telescope solidly and safely. If it's not properly set up, a mount can hurt you. Here are a few things to consider when you use a telescope mount.

4.4.3.1 Setting Up an Equatorial Mount

First, make sure the tripod or pier is solidly planted on the ground. Try to pick a level surface. If there are braces to keep the tripod legs from spreading, make sure they are attached. If the equatorial mount allows the polar axis to be adjusted for latitude, make sure it's tight before installing your telescope on the mount. If you're using a German Equatorial, do not fail to ensure that the counterweight is firmly attached to the declination axis. Ditto for the telescope.

4.4.3.2 Balancing an Equatorially Mounted Telescope

Your equatorially mounted telescope must be balanced around the declination axis, and everything rotating around the polar axis must be balanced, too. That prevents your drive motor from burning out by trying to move an unbalanced load against gravity. It also prevents the more delicate parts of your anatomy from being damaged by a suddenly moving counterweight shaft or tube.

Many words have been written about ways to balance a telescope. Some books go into this in great detail, so I'll just give you a few basics. These instructions apply mostly to balancing a German Equatorial Mount.

Balancing the tube—Position the telescope so the declination axis is perfectly horizontal. You can use a carpenter's level to verify it. Lock the polar axis brake. If the polar axis has a slip clutch, tighten it as much as you can. Unlock the declination axis brake. If it has a slip clutch, loosen it as much as possible. If the tube doesn't begin moving of its own accord, it is balanced. If one end begins moving down, that's the end too far from the intersection of the declination axis and the tube. Move the tube toward the other end. Keep adjusting the tube until there is no tendency for it to move under the influence of gravity. Relock the declination axis brake or retighten the clutch.

Figure 10: Types of Telescope Mounts

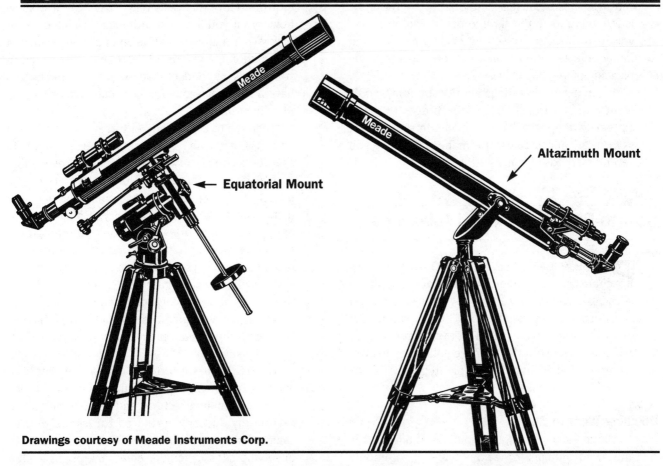

Equatorial Mount

Altazimuth Mount

Drawings courtesy of Meade Instruments Corp.

Balancing the polar axis—Release the polar axis brake or loosen the clutch. If the counterweight side of the axis wants to fall, move the counterweight closer to the axis until it's balanced. If it wants to rise, it needs to be farther from the axis. Move it away from the axis until it's balanced.

A *final reminder*—These instructions cover only the basics of balancing an equatorially mounted telescope. If you have a more complex telescope, with a camera or two strapped onto the tube, you need to check a book on telescopes. Or, ask a veteran amateur to help get you started.

4.4.3.3   Aligning the Polar Axis

When balancing an equatorial mount, you must align the polar axis so it points exactly at the celestial pole. Just ensuring that the polar axis appears to be pointing at the star Polaris when viewed from several different angles is usually adequate.

Photography and the use of setting circles require more accurate alignment. Many manufacturers now provide a small telescope mounted in the polar axis. It can be pointed at the pole through

a hole in the declination axis. A reticle is calibrated so the exact offset from Polaris can be made to ensure that the axis is pointing exactly at the pole. These mounts usually have reasonably clear instructions with them.

Fork-mounted Schmidt-Cassegrain telescopes do not include a telescope in the polar axis, because the tube would get in the way of seeing the pole. Many of these telescopes are equipped with finder telescopes featuring a reticle like the one in polar axis scopes. They also have instructions on how to align the polar axis.

4.4.3.4   About Setting Circles

Many equatorial telescopes are equipped with a circular dial on each axis that allows the telescope to be "blind-pointed" at a desired position on the Celestial Sphere. These dials are called "setting circles." The one on the polar axis measures the hours and minutes of right ascension, and the one on the declination axis measures degrees of declination. As you might suppose, setting circles require the polar axis to be almost perfectly aligned on the celestial pole.

I'm leaving out instructions on how to use setting circles because a main objective of this book is for you to learn the sky in order to find objects. If you learn how to use setting circles, you can find many objects "blind," but you'll deny yourself the pleasure of learning the sky.

If you get really serious about astronomy, you'll want to use circles to locate objects that are extremely dim, or to locate a planet in the daylight sky. An advanced telescope manual will give you several methods of using setting circles.

4.5
How to Purchase a Good Telescope

PLEASE, PLEASE, PLEASE, do NOT just read this chapter a few times, then go out and buy a telescope! You don't want to buy a telescope you'll never use, then banish the poor thing to your closet or the basement or attic. For the safety of your wallet and to avoid the telescope-abuse-by-neglect syndrome, follow these few simple steps:

4.5.1
Deciding What to Buy

I've already given you an outline of advantages and disadvantages for various types of commonly available telescopes. It's up to you to decide what you wish to observe (and/or photograph) in the sky, how much space you have to store a telescope, how much space you have in your vehicle to transport your telescope, how heavy a telescope you're willing to lift, and what portion of your spare cash you're willing to spend. A good general rule is: The easier a telescope is to transport, set up, and use, the more you'll use it. One exception: A telescope mounted permanently in some sort of observatory building that is easy to access will be used often, regardless of the telescope's complexity.

A final consideration I haven't yet covered is resale value. If you lose interest in astronomy or are transferred elsewhere by your company, can you get much of your purchase price out of your telescope?

4.5.2
Auditioning Telescopes

The local telescope store is NOT the place to audition a telescope. There are no stars to see. Go to the telescope store only to look at a bunch of tele- scopes. Make a card for each telescope you're inter- ested in, listing its features, and its pros and cons.

Join an astronomy club. (An enlightened propri- etor of a store that sells telescopes will also know how to put you in touch with your local club.) Your fellow club members will be an important source of information about which telescope is best suited to your needs and budget. A fellow member might also have a used telescope for sale that is close to your ideal instrument.

Then, go to a star party or three. If you've joined an astronomy club, you'll find such events are held as often as once or twice each month. Arrive before sunset so you can ask the proud owners of the telescopes being set up their opinions about their instruments. Take notes. Ask if you may come by after dark to view a few varied objects through the telescopes you would like to audition. After dark, take more notes. (Remember to use a very dim red- lens flashlight when writing notes after dark. Doing that will ensure your continued good health!)

Once you've narrowed your choices, discuss them with the veteran members of your club. Ask a lot of questions.

(NOTE: You should join an astronomy club whether or not you want to buy a telescope. Clubs provide informative lectures and demonstrations on astronomy and space-related topics, as well as camaraderie. Also, many clubs have observatories or observing sites in remote areas. By joining a club and sharing in the effort of building or maintaining an observing site, you can have most of the advan- tages of your own private site "out in the boonies" at a fraction of the cost.

Both major astronomy magazines in North America publish an annual supplement that lists clubs, planetariums, observatories, and manufac- turers of astronomical equipment. Since the maga- zines are competing to attract readers, they've made these lists, as well as their magazines, excellent.)

4.5.3
Where to Buy

While you might find the telescope of your dreams in the possession of a fellow member of your astronomy club, be sure to get another, more experienced member to look it over with you. Ask to set it up by yourself. See if it will fit in your car. Look through it before you buy!

Another source of used telescopes is classified advertising in the two major magazines, specialty classified advertising magazines, and in the classified ads in your local or regional newspaper. Check them all.

If you decide to buy a new telescope, you have a choice between purchasing one from a local retail store, often a camera shop, or purchasing an instrument via mail order. The local option has an obvious advantage: You can go to the store if you have a problem or need some advice on using your purchase. The disadvantage of dealing with a local merchant may well be that the price you pay will be considerably higher than via mail order, even if you add in shipping and handling charges. The local merchant also has a stock on hand for which he has already paid. He has some incentive to sell you what he has, rather than what you need or want.

The price is certainly right, but buying mail order can be tricky. There are many reputable mail-order merchants. There are about an equal number that are not.

Following are the attributes of a reputable firm, and those of a disreputable firm:

Attributes of a Reputable Mail-Order Dealer—Is willing to spend a few minutes on the phone or via mail to answer your questions...has been in business more than a few minutes...is an authorized dealer in the brand of equipment you wish to buy...clearly states all costs and clearly lists everything you're getting for the price you're quoted...when you call, has what you want in stock, or immediately available for drop-shipment from a distributor...has a clearly stated return policy you can live with.

Attributes of a Disreputable Mail-Order Dealer—Is vague about what you get for what price...has an enormous listing of products available, but, when you call, much of it is on "back order"...will not verify that a specific item is in stock...is unwilling to spend a reasonable amount of time answering all your questions...(Another sure tip-off: a vendor who wants to know what your credit card number is before he even asks what you want to buy.)...claims to be able to "beat any price, anywhere"...has an unbelievable return policy, or won't tell you what it is...will not accept credit card orders, or requires payment by money order only.

The last item is very important. If you pay with your major credit card, then find yourself in a dispute with a mail-order vendor, there are laws requiring the credit card company to become involved to arbitrate the dispute. If payment by money order is demanded, the vendor is forcing you to give up that consumer protection.

Do not assume from this discussion that I favor either local or mail-order purchase of a telescope. I've used both. I also have friends who operate both

types of businesses very reputably. Furthermore, some reputable manufacturers of specialty items sell them only through the mail because there is not a big enough local retail market for their product.

These are only guidelines, not a checklist, so they're not cast in concrete. But there are sharks out there, both locally and mail order. Beware!

4.6
A Telescope for Each Eye—Binoculars

Many people, myself among them, advise a beginner in amateur astronomy to begin looking at the sky through a pair of binoculars. Then, if the interest is there after a while, you can buy your first telescope. This section tells you what you need to know about choosing and using a pair of binoculars.

4.6.1
Why Binoculars?

There are several reasons to buy or use a pair of binoculars as your first optical aid in viewing the sky.

1. Binoculars are portable and easy to use, so you'll likely use them a lot. Even after you buy a telescope, a pair of binoculars will still be more portable than the telescope. You'll take them places you wouldn't take the telescope. They also have other uses besides astronomy, so your spouse will be less likely to hassle you over their purchase. (Whenever I go to a Dodgers game, there are no bad seats in the house. Also, using binoculars to watch a hawk glide through the sky in search of lunch is almost worth the price of the binoculars alone.)

2. Binoculars show a very wide field, so they provide a natural transition from the unaided eye to a telescope. Views of the Milky Way and other extended objects are exquisite through a good pair of binoculars. (When you get a telescope, they'll also serve as a finding aid.)

3. Binoculars also provide a natural transition in price between having nothing but this book and spending bigger bucks on a telescope. A satisfactory pair may be obtained for as little as $40. A quite good pair of binoculars may be obtained for as little as $100. (Whatever you do, buy the best binoculars you can afford, because lower-priced binoculars tend not to last. Also, after you gain experience, images that looked sharp when you started out will no longer satisfy

you. That's due to the ever-popular "training of the eye-brain combination.") Perfectly wonderful binoculars can be obtained for $300–$1000.

4. Since binoculars use both eyes, you see a bit more when viewing dim objects than you would if you viewed the same field through an equivalent low-power telescope. That's because any optical receiving device, including your eye, corrupts the view with some noise. Your eye-brain combination processes the two images it gets from the binoculars so that some of the noise is canceled, but the signal (the dim object) is not.

5. Since we spend most of our time viewing the world with both eyes, an instrument that allows us to view the sky with both seems to result in less eye strain. We see more.

4.6.2
Types of Binoculars

Whatever the layout of the optical system in binoculars, they are all characterized by a specification that, until you know the code, is a bit intimidating. Suppose you hold binoculars in your hand that are labeled "7×50 Field: 367 ft. at 1000 yards." The first number is the magnification, or seven power. The second number is the diameter of the objective lenses of the binocular, 50 mm (about 2 in.). The binoculars I've used as an example are commonly called "7-by-50s."

The part about the field was covered back in Section 2.5.2.2.3. The following table is repeated so you can determine the angular field of your binoculars:

Angular Field of Binoculars

Field (degrees)	Field (feet at 1000 yds.)
5°	262
7°	367 (Field you'll find on ZOOM maps)
10°	525
12°	631

There are basically two types of binoculars. Only one is useful to the astronomer. Make sure you know what you're being offered.

4.6.2.1   Field Glasses

This term describes binoculars that consist of two Galilean telescopes mounted side by side. A Galilean telescope uses a negative eye lens between the objective lens and its focal plane. This eye lens recollimates the cone of light (makes it parallel

again) and provides an erect image. The problem with a Galilean is that it is limited to very low power, 2× to 4×. To make up for that low power, it has a very narrow field. Field glasses of any quality are of little use in astronomy. They are better used as props in Civil War movies.

You can recognize (and avoid) field glasses by looking through them. If the view looks like you're at the end of a long tunnel, you're looking through a pair of field glasses. Put them out to pasture!

4.6.2.2   Prism Binoculars

Prism binoculars are more complex than field glasses. They consist of a pair of telescopes mounted side by side, each attached to a very stiff hinge. Rotating the halves of the binoculars around the hinge lets you vary the separation of the eyepieces to accommodate the separation of your eyes.

In more common types, a plate is attached to the hinge. The oculars (eyepieces) are mounted on the plate. When you turn a knurled ring, the plate moves in and out to focus the binoculars. To allow for the difference in focus of your two eyes, the right ocular of this type of prism binoculars also focuses independently of the ocular plate. This arrangement allows the binoculars to be focused quickly, but the ocular plate cannot be very thick, so it can have a tendency to flex. This flexure makes this type of binoculars difficult to focus accurately, especially in larger sizes.

More expensive prism binoculars dispense with the ocular plate, and have independently focusing oculars on each side, attached firmly to the body of each half. That allows these binoculars to be perfectly sealed against dust, dirt, and humidity. Many binoculars of this type are evacuated of air and water vapor during final assembly. Dry nitrogen gas is injected into the binoculars, and they are sealed. The dry nitrogen prevents the binoculars from fogging internally, and also prevents mildew from forming inside the tubes.

The prisms do two things: (1) They erect the image, by a process of flipping the cone of light coming from the objectives back on itself and around a few corners. (2) While erecting the image, the prisms also shorten the path between the objectives and their focal planes. That makes prism binoculars much more compact than field glasses.

As you might guess, keeping the prisms in just the right position is critical to ensuring that both halves of prism binoculars are pointing in exactly the same direction. How well the manufacturer allows for the precise alignment of the prisms in binoculars is one important factor that influences how much they cost.

Prisms in inexpensive binoculars are generally cemented in place in a special fixture that ensures the prisms are fairly well aligned. As long as you don't drop them they'll stay aligned. If you do, there's little you can do to fix them.

More expensive binoculars feature tiny screws so a technician can subtly move the prisms when aligning the binoculars. When new, these binoculars are generally better aligned than their less expensive cousins. If you drop them, they can be realigned by a technician.

Prism binoculars come in two types, one better for astronomy than the other.

Roof-prism binoculars have a complex, one-piece prism that erects the image without causing the light path to turn back on itself. They are no more compact than the same size field glasses. The roof prism is difficult to manufacture, but once it is made correctly, aligning the binoculars is very simple. However, roof prisms are expensive to manufacture in large sizes, so their use is limited to small binoculars.

Porro-prism binoculars are best for astronomy, because Porro prisms can be made economically in much larger sizes than roof prisms. Since their prisms can be larger, Porro-prism binoculars can have big objective lenses and can cover a wide field. Remember, you can tell if prism binoculars use Porro prisms if the objectives and the eyepieces are not lined up one behind the other.

4.6.3
Picking a Size to Buy

Since you're going to be using binoculars to view large, dim objects in the sky, picking a pair with an exit pupil of 7 mm will ensure that all the light the binoculars can intercept will enter the pupil of your eye. If you're a bit older, your dark-adapted eye may no longer open to 7 mm, so you may wish to select binoculars with an exit pupil of 6 mm or 5 mm.

Remember that dividing the diameter of the objective of any telescope by its magnifying power yields its exit pupil. Since binoculars are often used in low-light situations by military or law enforcement officers, manufacturers chose the nomenclature they use to make it easy for a prospective purchaser to determine the exit pupil. The 7×50 binoculars have a 7.1-mm exit pupil, because $50/7 = 7.1$. The 7×50 would be a good match to your eye if your dark-adapted pupil is about 7 mm in diameter. As our population ages,

7×42 sizes have become available, to provide an exit pupil of 6 mm. 7×35 binoculars have always been available, but they are usually made in that size to keep down size and cost.

Most amateur astronomers buy 7×50 binoculars if they are only going to purchase one pair. That size provides the brightest image that any binoculars can, and usually features a field of 7 degrees on the sky. Also, seven-power is the highest most people can hold steadily in their hands. The popularity of the "7 by 50" is the reason I've included a sample 7-degree-diameter circle with each of the ZOOM maps.

In recent years, larger binoculars have become more popular. You'll find 10×70, 11×80, even 14×100 sizes available. I use a pair of 10×70 binoculars almost exclusively. However, I usually use them while reclining in a lawn chair, because I can hold them steadier. Several years ago, I had the opportunity to view the summer Milky Way through a pair of 25×150 binoculars manufactured exclusively for astronomy. The quality of the views through those binoculars approached a religious experience. However, their price approached that of a mid-size imported automobile.

Zoom binoculars are attractive because you can vary the power over a moderate range. But only the most expensive zoom binoculars work reasonably well. They generally feature powers of 7×-14× or 7×-20×, so you'll not be able to hold them steady by hand. A large portion of the cost of manufacturing zoom binoculars is spent on the zoom optics. If you're going to spend a lot of money on a pair of binoculars, don't waste some of it on a capability you're unlikely to use. Buy fixed-power binoculars.

4.6.4
How to Look Through Binoculars

Many people use binoculars, but don't know how to focus them. Here's how to do it with center-focus binoculars:

1. Grasp the binoculars with both hands and look through them. Rotate the hinge until each of your eyes is exactly centered in an ocular.
2. Relax your eyes, and rotate the focus ring until the view through the left ocular is sharp. Because your eyes will differ slightly in focus, it's unlikely the right eye will be in focus.
3. Rotate the right ocular slowly until the view through that side of the binocular is sharp. If the left eye is a bit out of focus, repeat the process.
4. Once you've focused the binoculars, note the setting on the scale around the base of the right

ocular. That's the diopter scale, which measures the difference in power between your two eyes. Once you know what your diopter setting is, you can quickly set it on any binoculars and be fairly close to perfect focus.

Focusing individual-focus binoculars is a bit simpler. Once you've adjusted the hinge for the separation of your eyes, rotate each individual eyepiece until it is in focus. Memorize or record the diopter-scale settings on each ocular so you can quickly focus your binoculars the next time.

4.6.5

How to Buy a Pair of Binoculars

The comments in Section 4.5 also apply to purchasing binoculars. However, there are a few additional factors:

Since binoculars are used in fields other than astronomy, they are more widely available than astronomical telescopes, so the classified ads in your local paper are likelier to have a few more pairs of good binoculars for sale than astronomical telescopes. Sporting-goods shops and gun stores also carry a wide range of binoculars. [The ubiquitous garage sale will often yield broken binoculars. You can make a monocular by amputating the broken side with a hack saw. Such an instrument can be a wonderful (and inexpensive) gift for a kid-astronomer.]

Once you've found a pair of binoculars to buy, here's how to check them out:

1. Shake them. If you hear a "clunk-clunk" or a "tap-tap" when you do, a prism is loose. Put these out to pasture in the field-glasses field.
2. Look into each objective. Do you see any chips, dirt, or scratches? If so, these are a pair of "rejectoids," too.
3. Hold the binoculars at arm's length and look at the exit pupils. Do they look round? If so, the prisms are large enough to allow all the light coming from the objective to reach your eye. Do they look square? If so, the manufacturer used small prisms, so much of the light coming from the objectives will never reach your eye, resulting in a dimmer image than with binoculars with fat prisms. These "square-oids" are also "rejectoids."
4. Look at a distant object, preferably some electrical wires silhouetted against the sky. Relax your eyes and focus the binoculars carefully. Do both images line up perfectly? Do you have to strain a bit to get them to line up? If so, these

are "rejectoids," too, because their prisms are a bit out of alignment. (You will see a faint violet "ghost" image at the edge of a dark object silhouetted against the sky. That's normal, because most binoculars have achromatic objectives, which only focus two colors in the same plane. The other colors are close. Violet is the color to which our eyes are least sensitive, and its image is farthest from the plane of the others, so it looks bigger. The violet image is the "ghost.")
5. Look at the surface of the objectives and the eye lenses. If they are distinctly amber or violet in color, the binoculars feature antireflection coatings, which suppress reflection of light from the surfaces of the lenses and prisms. Since this stray reflection is suppressed, there will be higher contrast, and the view through the binoculars will be enhanced. Binoculars labeled "multicoated" feature an enhanced antireflection coating, which results in stray light being absolutely eliminated. (Reject uncoated binoculars.)

If a pair of binoculars you are auditioning passes these five tests and if you can afford them, these binoculars should perform well when you look at the sky through them. Enjoy!

4.7

Quick Notes on Photographing the Sky

Although it is very sensitive, your eye only stores light for 0.10 second before it begins another "exposure" of an image to send to your brain. If you could store light for seconds, minutes, or hours, the light from faint objects in the sky would build up and you could see incredibly faint objects.

Photography, of course, allows you to build up an image on film. Astrophotography has become one of the most popular and interesting activities in which amateur astronomers participate. Sometimes, using only modest equipment, amateur astrophotographers capture breathtaking images of the stars, deep-sky objects, the Sun, Moon, and planets.

Amateurs have been taking photographs of astronomical objects since the twenties. Until the last ten years or so, the recording medium has been photographic film. But electronic imaging has become more and more popular in recent years.

This book is devoted to helping you see the sky, so I'll not devote any more space to photography. For more, see a book devoted to astrophotography.

CHAPTER 5

Picking a Time and Place for Viewing

The Astronomer's Worst Enemy— Earth's Atmosphere

Astronomers have a love-hate relationship with Earth's atmosphere. We love it because we're fond of breathing, and we hate it because even when the sky is clear of clouds, it limits how well we see the sky. In fact, the atmosphere is almost always the limiting factor on what we can see through our telescopes. Here's what's happening:

5.1.1

About Turbulence and Seeing

Few terms in astronomy are misused more often than "seeing." That would be O.K., except the person who's misusing the word will also not know what, if anything, she or he can do to improve the view. Pay close attention to this section.

5.1.1.1 Earth's Atmosphere Refracts the Light from the Stars. How Much Is a Measure of the Seeing

When refraction was covered in Chapter 4, I wrote that light is bent, or refracted, when it passes through a transparent material like glass. Water bends light almost as much as glass, while air bends light only a tiny bit, less than one-thousandth as much as water.

Much of the time, air being refracted by the atmosphere causes astronomers a lot of grief because the pressure of the air through which light passes determines the amount the light is refracted.

If something only slightly changes the pressure of air, the light from a star passing through it is refracted only slightly. The star appears to have moved a bit, because the path it took before it was intercepted by your telescope was deviated by the change in pressure.

If this pressure change is slow enough, your eye just follows the star image as it moves in your field of view.

As you might imagine, if the pressure changes in the atmosphere were either very great or slight, but very fast, your eye would have great difficulty in following the star. You would see a blur. If the star were quite dim, its light might be spread out over many of the sensors in your eye and they might not tell your brain they see anything at all.

How little or how slowly the atmosphere is moving the image of a star around is a measure of how steady the atmosphere is between you and the star. Although I prefer the term "steadiness," the term used by most astronomers is "seeing." So we'll use it, too. Bad seeing is caused by turbulence in the air somewhere in the path starlight must take before it gets to your eye.

5.1.1.2 Stirring Up the Pot—Turbulence

To get an idea of what turbulence is, consider a waterfall. As water flows over the edge, the flow is quite smooth, even though it is moving very fast. This smooth water is called "laminar flow," because you could think of it as being made up of many thin layers, or *laminates*. If you could take a snapshot of a laminar flow, and slice it vertically, you might see each laminate become thicker or thinner, but you would see that change occur smoothly and gradually.

Light passing through a fluid in laminar flow might be refracted a lot, but its path through the area of laminar flow would not change much over time. The image of a star seen through a fluid in laminar flow would be sharp. How sharp depends on just how smooth the laminar flow is.

As the water flows down the waterfall, the smoothness of the flow decreases fairly gradually, until at the bottom the flow is in utter chaos. Just below the edge, we might still see large objects. As the flow becomes rougher, paths of light through the water become more refracted, so we see less and less. At some point, nothing can be seen, because the flow becomes lost in chaos.

This rough flow is called turbulence. Fortunately, the seeing in Earth's atmosphere is not nearly so bad as the seeing in a waterfall. However, turbulence can and often does place severe limits on what we can see with our telescopes.

5.1.1.3   About the "Seeing Disc"

Since turbulence is random, light from a star passing through turbulence is spread over a circle, called the "circle of confusion." Many others call that the "seeing disc," so that's what we'll call it. The diameter of the disc is what astronomers use to rate the seeing. Here's a Seeing-Rating Table:

Seeing-Rating Table	
RATING	SIZE of SEEING DISC (secs.)
Excellent	0.5" or smaller
Good	0.5–1.0"
Fair	1.0–1.5"
Poor	1.5–2.0"
Hopeless	Bigger than 2"

Keep several factors in mind when you think about the seeing disc:

1. If the diffraction disc is bigger than the seeing disc, you won't be able to tell what the quality of the seeing is. You simply won't be able to resolve it. So, the smaller your telescope, the less often you'll be affected by bad seeing. For instance, a 60-mm refractor can only resolve two bright stars that are about 2" apart in the sky. The best 60-mm refractor, then, could only tell the difference between what I've labeled "poor" and "hopeless" seeing.

2. Conversely, the larger your telescope, the more often it will be affected by seeing. One primary reason is that it's more likely that your larger telescope can resolve the seeing disc. Also, since your larger telescope can intercept much more light than a smaller one, you're more liable to see any light that has been spread around by the turbulence.

5.1.1.4   Turbulence and "Twinkling"

By now you may have figured out that turbulence is the cause of stars twinkling. You're right. However, the techno-term for "twinkling" is scintillation. If you once loved to watch the stars twinkle, as soon as you buy a telescope and point it at a planet, you'll change your mind. If you can see a star twinkling with your unaided eye, which resolves poorly, think what the size of the seeing disc will be when you look at that star through your telescope.

Except when scintillation is extreme, we don't see planets twinkle. That's because the angular size of a planet's disc is almost always much greater than the seeing disc. Bad seeing just spreads the planet's disc a bit, so we don't see it appear to flicker the way stars do. If you see a bright "star" in the sky, and it's not twinkling along with the others, it's very likely a planet.

5.1.2
A Clean Window vs. a Dirty One—Transparency

I've probably convinced you that seeing is ultra-important to viewing the sky in a telescope. That's true. However, there is another factor, transparency, that is also important. As you might imagine, transparency refers to how much of the light from the stars gets through the atmosphere to be intercepted by your telescope's objective. However, it's a bit more complicated.

5.1.2.1   About Signals and Noise—Scattering

In Section 4.3.3.2, you learned the effect diffraction has on contrast. In that context, I was referring to light being scattered inside your telescope. When we talk about sky transparency, we're talking about light being scattered in Earth's atmosphere.

Scattering occurs when a ray of light is intercepted by a molecule of gas or a particle in the atmosphere, and its path is bent by the collision. Unlike refraction, a very predictable phenomenon, scattering is random. When light is being refracted, its path can be bent only one way. The direction in which light will be scattered is completely random.

In that context, think of the light that is merely refracted by the atmosphere and then intercepted by your eye as the signal, the message you want your eye to receive.

Some light that's been scattered by the atmosphere also enters your eye. Since this scattered light is random, you perceive it as a haze. You can think of that haze as "noise" that's interfering with your ability to see all the unscattered light reaching your eye.

5.1.2.2   How Scattering Reduces Contrast

Remember what I wrote about contrast in Section 4.3.3.2 when I introduced the concept of contrast as the ratio of the brightest to the darkest, and said that perfect contrast, or 100 percent contrast, was achieved by placing a pure white object next to a pure black object.

In the previous section, it was signal vs. noise, where the signal was the light from an interesting celestial object you want to enter your eye, and noise is the light that's scattered by the atmosphere, which you don't want to enter your eye.

Electronics buffs will recognize the concept of "signal-to-noise ratio" or SNR, which indicates how much greater the signal is than the noise found in any electronic device. To the astronomer, contrast is the SNR of the view.

As scattered light increases, an object must be brighter and brighter to be distinguished "above" the noise. In the big city, where there is an abundance of artificial light being scattered by the atmosphere, the SNR may be so low you cannot see anything but the brightest stars. If you get away from the city, the contrast isn't perfect, but it's much better.

5.1.2.3 What Does the Scattering and Reduces Transparency?

We can blame four main culprits for loss of transparency. Three we can minimize, the fourth we can't really do much about, unless we leave Earth altogether.

5.1.2.3.1 The Gases in Our Atmosphere Scatter Light Proportional to Its Wavelength

Our atmosphere is 79 percent nitrogen, 19 percent oxygen, and 2 percent "other." These gases do all of the light refracting, and some of the scattering. But there's one significant difference between what the gases scatter and what other constituents of the atmosphere scatter. The gases scatter light proportional to wavelength. The shortest wavelengths we can see, blue and violet, are scattered about sixteen times as much as the longest, the reds.

Since scattered light is mostly blue, with a diminishing mixture of green, yellow, orange, and red, we see the daytime sky as blue.

The only way we can minimize this form of scattering is to go to a high altitude, where there is less atmosphere to scatter the light of the stars. When we do go to a high altitude, we see the sky as a deep blue-violet. That's partially a result of thinner atmosphere, and less scattering.

5.1.2.3.2 Water Vapor Is an Equal-Opportunity Light-Scatterer

Water vapor is responsible for the ever-popular saying "It ain't the heat; it's the humidity!" Humidity does more to astronomers than make them sweat. Water vapor in the air tends to scatter all different colors of light equally. As the humidity rises, a deep blue sky doesn't change color, it gets lighter and lighter, as more and more white scattered light is mixed with it. In general, the bluer the daytime sky, the lower the humidity.

If you live in an area where humidity is high, you know that even on a clear day, you can't see forever. Instead, you notice that beyond a certain distance, detail is lost in a fog of gray. If you travel to a desert area, you may find that you can see many times farther than you can at home. In the desert, it is not uncommon to be able to see 100 miles or more before the little humidity in the air reduces the contrast enough to blot things out.

How much water vapor can remain mixed with the gases in our atmosphere depends both on altitude and on the temperature of the atmosphere. The higher the altitude (and, thus, the lower the atmospheric pressure), the lower the humidity. Also, the lower the temperature, the lower the humidity. Observing the sky from a high mountain on a cold night ensures a very transparent sky.

5.1.2.3.3 Smoke and Dust Like to Scatter Light

Smoke can run the gamut from a thin cloud of ultrafine particles to dense sooty particles. If smoke is mostly fine particles, it tends to scatter more blue light than the longer wavelengths of light, much as the gases in the atmosphere. When smoke is made up of larger particles, it tends to scatter light of all wavelengths more equally.

Dust in the atmosphere also tends to scatter light equally with respect to wavelength, so it behaves a lot like water vapor.

When the air is humid, water vapor tends to condense onto particles of smoke and dust, creating a much larger particle to scatter light. This composite particle is what does a lot of light scattering when humidity is high.

5.1.2.3.4 Volcanic Aerosols REALLY Scatter Light

You can travel way out into the country or to the top of a mountain to escape light pollution caused by a combination of humidity, dust, and excessive light in an urban area, but if a major volcanic eruption occurs, the transparency of the sky over much of our planet is badly affected.

Many people assume the cause for transparency reduction is volcanic dust blasted into the upper atmosphere by a major eruption. But that's only a small part of the problem.

The real culprit is the huge dose of aerosols blasted into the upper atmosphere. Chiefly sulfates or ions containing the element Sulphur, these sulfates mix with water vapor to form sulphuric acid, the same corrosive stuff that causes acid rain.

A sulphuric acid molecule is much larger than a water molecule, so it is much better at scattering light. Since it's so big, it tends to scatter a lot of light, and does so almost uniformly; it behaves like a super water molecule.

Because the volcanic aerosols are blasted very high into the atmosphere, it takes a long time for them to work their way to the lower altitudes where they are washed out as acid rain. Consequently, for many months after a major eruption, the night sky is not nearly so dark as it was before an eruption.

Volcanic eruptions like those of Mount Pinatubo in the Philippines and El Chinchonal in Mexico wrought havoc in the night sky, seriously increasing light scattering and reducing atmospheric transparency. However, that pales in comparison to the misery those eruptions caused to the people whose bad fortune it was to live too close to those two mountains. We should count our blessings.

5.1.2.3.5 Thumbing Your Nose at the Sun (to Test for Atmospheric Scattering)

You can make a pretty good guess how much light the atmosphere is scattering simply by holding your thumb at arm's length so it eclipses the Sun from your view. If there is a significant white halo around your thumb, there are two possibilities: (1) Your thumb is eligible for sainthood, or (2) there's a lot of scattering of the Sun's light by the atmosphere. The smaller the halo of light around your thumb, the better the transparency should be that night. Remember, though, that this test can't account for a change in the weather between when you perform the test and later that night. (This test also works fairly well when the Moon is near its full phase.)

5.2

When and Where to View What

5.2.1

When to View What

Although I may not have the enthusiasm of a cheerleader, I must confess I've never met a view of the sky I didn't like. However, some were better than others. So, I look at the sky when I can and urge you to do the same. The more you look, the more you'll see.

Here are some general guidelines that will maximize your enjoyment:

1. The later at night you view, the likelier it is that the seeing will be steady, because the Earth spends the first part of the night giving off the heat from the Sun it absorbed during the day. Part of the heat is radiated into the sky as infrared light, but quite a bit more flows into the atmosphere through convection, which creates turbulence. Some of the very best seeing I've ever encountered was just before sunrise. Since the Sun takes some time to warm things up after rising, the seeing will often continue to be good after sunrise. I've observed the bright planets well after sunrise with considerable success. Try it.

2. Since transparency often improves as the atmosphere "dries out," it should improve as the night grows colder, because cold air cannot hold as much water vapor as warm air. The price you pay for the atmosphere's shedding its water vapor is dew, which has a nasty habit of getting into everything, including your optics.

3. When a cold front passes through your area, it is usually accompanied by rain or snowstorms. Storms almost always wash dust from the atmosphere. The cold air that rushes in after the front passes can't hold much water vapor so scattered light will be minimal, since there's less stuff in the atmosphere to scatter it. However, because there's a lot of cold air to create turbulence, seeing will generally be poor the first night after a front.

   In general, the night after a cold front passes through your area will feature very good transparency, and poor seeing. If you're out observing on such a night, expect to view many large, dim objects at low power. Save your high-power views for nights when the seeing is better.

4. The last clear night before a cold front is expected, or the night before, will generally be the one when you expect the best seeing. Remember that "later is better" guideline. (If you receive the Weather Channel, you can track the progress of storms as they approach your area. If that channel is not available, try either CNN or CNN Headline News, both of which feature frequent weather reports, the latter every thirty minutes.)

5. It's ironic, but if you live in a city that suffers from smog, you can use the smog to tell you when to view the planets and other bright objects that you would view at high power. The presence of smog indicates stagnant air, the

opposite of turbulence. If you find yourself suffering a particularly smoggy day, you may find that the early evening will supply spectacular views of the planets.

6. In general, the higher an object is above the horizon, the less its image will be affected by the atmosphere. That applies both to seeing and to transparency. The reason is simple: The less atmosphere the light from the object must traverse, the less turbulence it will encounter and the less it will be scattered by molecules, water vapor, smoke, dust, and volcanic aerosols.

This means that if you want to ensure the best views of an object, observe it as it crosses the meridian. That's when it is at its highest altitude above the horizon.

5.2.2
Where to View What

Earth's atmosphere is incredibly complex, so finding a place from which to view the sky is often a matter of chance. Just as with the issue of when to view, however, there are some guidelines you should follow in picking a place from which to view.

1. Stay away from places that are sources of heat. Sure, it's convenient to set up your telescope in a nice parking lot. However, the enormous mass of concrete or asphalt from which the lot is made absorbs a lot of heat from the Sun during the day and serves as a wonderful source of heat and turbulence once the Sun sets. Other locations to avoid are: major highways, airports, large concentrations of buildings, power plants in general, nuclear power plants in particular, stadiums, and manufacturing plants.

If you're observing from home, just placing your telescope so you're not looking over your roof or your neighbor's will minimize bad seeing.

If you travel to an observing site, be sure to put your car as far from your telescope as you can. Placing it to the north is the best choice, as you're less likely to be viewing in that direction.

If you're observing with a group of people, make sure they stand or sit to the north also, rather than being right under the end of the telescope. If you doubt the wisdom of this, consider that each human being emits the same heat as a 60-watt light bulb. Do you want that

much heat flowing through the atmosphere just in front of your telescope?

Your telescope itself is a source of heat. If you take it out of a warmed room into the cold night air, then begin viewing through it immediately, you won't like what you see. That's because currents of warm air are being generated by your telescope. Until they carry off its heat, they will cause turbulence and bad seeing. Allow thirty minutes to an hour for your telescope to reach equilibrium with its surroundings. Better yet, store it in an unheated building so it stores less heat.

2. Avoid valleys, because as the Earth cools, the air above it cools and cold air tends to collect in valleys. There will often be a layer of warm air over the cool air in the valley. This is called an inversion layer, because the change in temperature is inverted from what happens during the day. There's often a lot of turbulence just below the inversion layer, but it's often quite calm above the inversion layer. If you can, climb a hill to get above that layer.

3. One place that might feature good seeing would be the top of a hill that's above the inversion layer, with a nice forest of evergreens growing on its slopes. The trees are important because they spend the day absorbing much of the Sun's energy and converting it into more tree. (The techno-term for this is "photosynthesis.") Since they permanently convert much of the Sun's energy they receive into more tree, they don't give off so much of it into the atmosphere at night, reducing the possibility of turbulence.

4. It's probably obvious that to see dim objects, you must observe the sky from a place in the boonies where artificial lights are few. If you can combine that spot with the pine-covered hill, so much the better.

As state and national parks become crowded, taking your telescope to one of those loses appeal. You might try asking the owner of some land in a rural area for permission to set up your telescope in his pasture. Private camps and retreats operated by church and other nonprofit organizations are another source of places from which to observe the sky. Wherever you go, be sure to offer views through your telescope in exchange for use of the site.

CHAPTER 6

Looking at the Sky After Sunset

Even if you care little for the scientific explanations for what you're seeing, observing the sky just after the Sun sets can be most rewarding. The heat of the day is abating. It's a bit quieter than during the hustle and bustle of the day. Just sitting in a comfortable reclining lawn chair and watching day give itself up to night can be an important source of energy for recharging your personal batteries.

If you're traveling to a remote site to do some sky watching, try to arrive well before sunset, so you can set up your telescope and have an unhurried bite to eat before sunset. Then you can relax and enjoy the subtle but beautiful change from day to night without hurrying to get ready.

And, of course, for every sunset, there is a matching sunrise. You can see everything I'll describe in this section either after sunset or before sunrise.

6.1

A Bit More About Scattering

We know why the sky is blue, but why is it increasingly red after sunset? And why is the Sun itself very red when we see it on the horizon just before sunset?

Both effects are caused by scattering of the Sun's light by the atmosphere. Since the Sun's light must pass through many times more atmosphere at sunset than when the Sun is high in the sky, almost all the blue portion of the solar spectrum is scattered, less green, less yellow than green, even less red. Not a lot of the red gets through, either, but since there's more red than any other color, the sky looks red at sunset.

6.2

Watching the Sun as It Sets

Most of the time, there is enough crud in our atmosphere so you can observe sunset without damaging your eyes. A general rule: If you have to squint, don't look at the Sun as it sets.

As you watch the red Sun set, you'll probably notice that the Sun appears to be flattened, as if William "Fridge" Perry were sitting on it. It looks that way because our atmosphere refracts the light from the Sun. Since rays from the Sun are bent down toward us by the atmosphere, we can actually see the Sun after it is physically below our horizon. That means a bit more than half of the Celestial Sphere is being squeezed into the hemisphere above our horizon than we would see if refraction wasn't happening. Since the Sun is being squeezed from below the horizon to above it, its image is squeezed vertically, but not horizontally. When the Sun has finally set, it is actually a bit more than its own angular diameter below the horizon.

Refraction also plays tricks on the last image of the Sun as it sets, so viewing the Sun with binoculars as it sets can be rewarding. Sometimes the turbulence through which the light from the setting Sun must pass is so great the image breaks into several distinct layers. At other times, when there is little turbulence, the last image of the Sun will be green. This "green flash" happens very quickly, because green light is bent a bit more than red. When the "red image" of the Sun has set, the "green image" is still up. Images corresponding to even shorter wavelengths might exist, but many of their photons are scattered by the atmosphere, so they are too dim for us to see. This phenomenon is seen very rarely, mostly because people don't look for it But if you look at a lot of sunsets, eventually you'll see the green flash.

6.3

There's Twilight, and Then There's *Twilight*

The farther the Sun is below the horizon, the less its light is scattered by the upper atmosphere into the lower atmosphere, and thus, the darker the sky will be. There are three definitions of when twilight ends after sunset or when it begins before sunrise. All three are determined by the angular distance of the Sun below the horizon. The distance for civil twilight is 6 degrees, for nautical twilight, 12 degrees. Unless you combine astronomy with the law of the sea, astronomical twilight is the definition that will appeal to you. For astronomical twilight, it's 18°.

If you measured the average brightness of the sky at the zenith from sunset to sunrise, you would see it started very bright, then tapered off until it reached a constant very low level, at which it remained for most of the rest of the night, until it began to brighten again before sunrise. The points where the curve of the sky brightness "flattens out" is where the Sun is 18 degrees below the horizon.

If you lived at the equator, the Sun would appear to plunge "straight down" at the western horizon at sunset and rise "straight up" from the eastern horizon. There, the length of time between sunset and the end of astronomical twilight would be minimal, because the Sun is moving "straight down." It would, in fact, be only 72 minutes long. If you've ever traveled to a place near the Equator, you've no doubt noticed it gets dark quickly there after sunset.

As you move farther and farther north, the angle the Sun's path makes through the sky with the horizon gets flatter and flatter, so it takes longer and longer for the Sun to get 18 degrees below the horizon. The farther north you travel, the longer astronomical twilight lasts and the more the length of the completely dark night shrinks.

At north latitude 50 degrees, there's a five-week period centered around the summer solstice when the Sun never gets 18 degrees below the horizon, so the nights are not completely dark. By the time you get to 66 degrees north, the period of continuous twilight lasts from early April to early September. Of course, farther north than 66.5 degrees, the Sun never sets at all for some days near the time of the summer solstice.

If you live very far north, on nights in the middle of summer you'll find that you will not get particularly good views of the dimmer deep-sky objects, because twilight will "wash them out." However, the Moon, planets, and double stars will be unaffected.

6.4

Seeing the Earth's Shadow

If you're observing the sky just after sunset or just before sunrise, be sure to examine the horizon just opposite where the Sun has set or will rise. About 20 to 30 minutes after sunset, as the sky begins to darken appreciably, you should be able to see a dark purple band on the horizon opposite where the Sun has set. As you watch, the band will slowly rise, until the sky darkens so much you cannot see it any longer.

What you're seeing is the shadow of the Earth projected onto the atmosphere. Above the line, just a bit of direct (mostly red) light is hitting the atmosphere. Below the line, only scattered light, which has a distinctly bluish color, is reflecting off the atmosphere back to your eye.

This effect is not at all obvious, so to see it you must be well away from the city. Since humidity scatters a lot of light, you're less likely to see the Earth's shadow from a place where it's humid than from a dry place.

Although it's not a spectacular sight, many people are attracted by the subtle beauty of the Earth's shadow. Whenever I'm out in the mountains observing the sky, I make it my first observation.

6.5

Finding the First Stars and Planets (and a Daytime Surprise)

Although of no particular scientific value, finding the first stars and planets visible after sunset can help you sharpen your visual acuity. Doing so with a group of kids can also be a fun activity. However, be prepared to confront the fact that kids can see more than you can, since their eyes have a wider "bandwidth," that is, their eyes can see farther toward either end of the spectrum than adults'. (That's why we remember that colors looked more vivid to us when we were kids than they do when we're older. Because they were!)

You can use the methods in Chapter 3 to determine where bright stars and planets are in the twilight sky. After a few tries, you'll find you're seeing the first star or planet earlier and earlier after sunset.

Several planets can be seen during daylight, despite the fact that the Sun's light is being scattered by the atmosphere. Since it is the brightest, Venus is the easiest planet to spot in the daytime sky. Both Jupiter and Mars can also be quite easy, depending how bright they happen to be. Although I've never observed Saturn during the daytime with the naked eye, I've observed it several times with a telescope.

As you might expect, the better the transparency of the sky, the easier it will be to see a planet in daylight. Really humid days will offer you little chance to spot a planet when the Sun is up. Sometimes, the humidity can be so high you can't even see the Moon in daylight!

Morning is better for daylight planet spotting than afternoon, because the Sun has not yet warmed the atmosphere enough to cause turbulence that would pick up dust to obscure your view. Also, the temperature is generally lower in the morning than in the afternoon, which means the atmosphere can't hold as much water vapor to scatter light as it could in the warmer afternoon.

Once you find Venus in the daytime sky, you'll realize it's not particularly hard. "Why haven't I seen that before?" you might ask. Easy. You never looked! That is true for many things in the sky. All you really have to do is look.

6.6
Seeing the Zodiacal Light

When you go away from the city to observe the sky, it's a good idea to know in which direction all the nearby cities lie from your observing site. That's so you'll know in which direction there will be a glow near the horizon caused by the atmosphere scattering the glut of light from the cities. Then you can plan to observe only those very dim objects that are not affected by light pollution.

Another reason to know where to expect a horizon glow caused by the presence of a city is that there is one glow not caused by anything on Earth. It's called the *zodiacal light*.

You'll see the zodiacal light as a wedge-shaped brightening of the sky at either the western horizon after sunset or the eastern horizon before sunrise. The base of the wedge is at the horizon and can be as wide as 30 degrees. The middle of the base marks where the ecliptic meets the horizon. The vertex, or point of the wedge, lies on the ecliptic, and can be as much as 90 degrees from the base.

Astronomers believe the zodiacal light is caused by light from the Sun being scattered by grains of dust found in the plane of Earth's orbit. The light is most intense near the horizon because light from the Sun is hitting the grains of dust at a grazing angle, which is the most efficient way of reflecting the light. The grains farther away from the Sun in the sky scatter light more randomly, so not as much is reflected to your eye.

On very rare occasions, experienced observers have seen a uniform band of light continue from the vertex of the zodiacal light along the ecliptic to the opposite horizon. This is called the *zodiacal band*. Approximately 10 degrees wide, it is considerably dimmer than the zodiacal light.

At the point on the ecliptic just opposite the Sun, the zodiacal band can be seen to widen into an approximately circular patch 20–25 degrees wide. This is the *gegenschein*, "counterglow" in German.

As with the zodiacal light, astronomers believe the zodiacal band and the gegenschein are caused by scattering of light from the Sun off of dust grains. However, some astronomers believe the gegenschein is caused by reflection of the Sun's light from larger particles as far away as the asteroid belt between the orbits of Mars and Jupiter.

In the temperate latitudes of the northern hemisphere, the zodiacal light is best seen near the western horizon on late-winter and spring nights, just after astronomical twilight ends. Under those conditions, the ecliptic makes the least flat angle with the horizon, so the light is not lost in the mist near the horizon. (Remember, this is also why evening apparitions of Mercury are best seen in late winter and spring.) Seeing the zodiacal light in the morning is best done in the late summer and fall.

The zodiacal light is about as bright as the Milky Way, so even though it is near the horizon, it is relatively easy to see from dark, dry locations. I've seen it many times.

I've only seen the zodiacal band and the gegenschein exactly once, at the Lockwood Valley, California, observing facility of the Polaris Observatory Association, when my friend Steve Edberg pointed them out to me one clear, dry, and very dark night in March several years ago. I suspect I could have seen them on several occasions, but because I had always read that they were "almost impossible" to see, I never looked for them. Steve cured me of that. As a beginning observer, it is very unlikely that you will see these two phenomena. If you don't ever look for them, though, it is a certainty you won't see them.

6.7
Seeing Artificial Satellites

Recently I read in the *Los Angeles Times* that the North American Air Defense Command keeps track of approximately 8000 separate and distinct objects orbiting our planet. These consist of spent booster rockets, shrouds, other miscellaneous debris, and, of course, the payloads themselves.

Most of the 8000-odd artificial satellites are so small only sophisticated radar devices like NORAD's can detect them. (And, it's a good thing they do, because the Shuttle must occasionally dodge a piece of debris in order to prevent damage to its protective heat-shield tiles.) However, there are objects in orbit you can see with only your eyes. It's not hard, but you must look from a reasonably dark site and know when and where to look.

When is between the end of evening twilight and approximately an hour later. During that period, there will be satellites in direct sunlight above you. They will reflect the sunlight down to you and will appear to be like a third- or fourth-magnitude star, except, of course, they're moving. Don't bother to try to see the shape of a satellite through your telescope. They're all too small for that.

Where is the half of the sky nearest the Sun, as that is where the satellites have the best chance of still being in direct sunlight. Since almost all the objects you're likely to see are reconnaissance satellites, you'll see them moving in polar orbits, so they'll be moving in a generally north-south or south-north direction.

When you spot a satellite, follow it with your binoculars or finder telescope. If it moves generally away from the Sun, at some point you are liable to see it begin to fade and brighten as it goes into the Earth's shadow. If you see several satellites in a brief period, you can gauge where the edge of the shadow is by noting where the satellites "go out."

If you see an object moving at satellite speed, but brighter than any star, that's Mir, the space station of the former Soviet Union. It's so big astronomers at an observatory in Arizona were able to photograph it with a 90-in. telescope. The photo showed its main body and solar power panels. Other very experienced satellite observers have, literally, been able to observe Mir as the Russkies threw their garbage over the side. Really!

If you can get information from your local newspaper or another source, the Space Shuttle is also easy to see. How bright it will appear depends on how its pilot has oriented it as it orbits. However, since it's about the size of an MD-80 twin jetliner, it will always be bright, though not quite as bright as Mir.

The next time you're out in the boonies with your friends, be sure to look for a satellite or two after sunset. You can dazzle 'em with the satellites' brilliance, rather than trying to baffle 'em with organic material.

6.8
Seeing a Very Young Moon

Amateur astronomers make great sport of trying to see a very young Moon. That just means they try to see the Moon in the evening sky as soon after new moon as they can.

Since the Moon is an extremely thin crescent when it is less than 24 hours old, it is also exceptionally dim. Compounding the problem of seeing it is that it cannot be seen against a dark sky, as it will set before twilight ends. Finally, the light from this dim crescent must pass through a "stack" of the Earth's atmosphere, which will dim the light some more. (It will also redden it.)

All this makes observing an extremely young Moon a real challenge. Here are some tips on how to do it:

1. Observe from as high and dry a place as possible, which will minimize the amount of atmospheric crud that might spoil your view.
2. Plan your observation carefully. Know exactly where the Moon is supposed to be. You can do that by noting where the Sun set. The Moon cannot be more than five degrees to the north or south of that point.
3. Don't try to find a young Moon with the naked eye. Use binoculars to slowly sweep the area where the Moon must be.
4. Finally, know that finding a very young Moon, like finding Mercury, is easiest in late winter and spring, when the ecliptic makes the least flat angle with the western horizon after sunset.

6.9
Seeing the Aurora

Occasionally, energetic charged particles are flung off the Sun into the solar wind. When they reach the Earth, our planet's magnetic field attracts them to the polar regions, where they congregate in the upper atmosphere, and excite various atoms into emitting light. In the northern hemisphere, this phenomenon is called the Aurora Borealis. A more popular term is the Northern Lights. In the southern hemisphere, they're called the Aurora Australis.

The Northern Lights can be spectacular, multi-colored displays, covering huge portions of the sky. Only in rare instances will you see the Northern Lights from the temperate latitudes. However, the farther north you live, the more likely you'll see them.

CHAPTER 7

Inside the Solar System—the Sun, Moon, and Planets

Although the previous chapter gave you information on what to observe in the sky, this is the first chapter devoted to observing the more classical objects in the sky. Let's begin with those objects we humans have observed since the beginning of time: the Sun, Moon, and planets.

Each section devoted to a particular type of object stresses how to observe that object. I assume your curiosity demands that you know a bit about what makes each type of object tick, so I've also included more on that.

7.1

General Description of the Solar System

Earth's immediate neighborhood, the Solar System, is dominated by the nearest star, the Sun. There are many bodies orbiting the Sun in egg-shaped orbits called ellipses, though many of these orbits are nearly circular. Others are so elliptical they look like thrown javelins. All the planets and most of the other objects orbit the Sun in one direction. They are said to be in direct orbits. A few oddballs are in retrograde orbits.

Besides the nine planets, there are asteroids, comets, dust, gas, charged particles, other particles for which God paid cash, and a few artificial spacecraft.

Because the Sun is radically more massive than everything else in the Solar System put together, it dominates the system. Except that we live on one of the planets, we might well describe the Solar System as the Sun*. That asterisk would cover everything else in a small footnote at the bottom of the page. Illustrating this point is the fact that the Sun has about 740 times more mass than everything else in the Solar System.

In theory, the Sun and its planets all formed about five billion years ago out of a vast, lens-shaped mass of hydrogen gas and dust. Most of the mass condensed into the Sun, while a few other large

Figure 11: What You're Gonna See

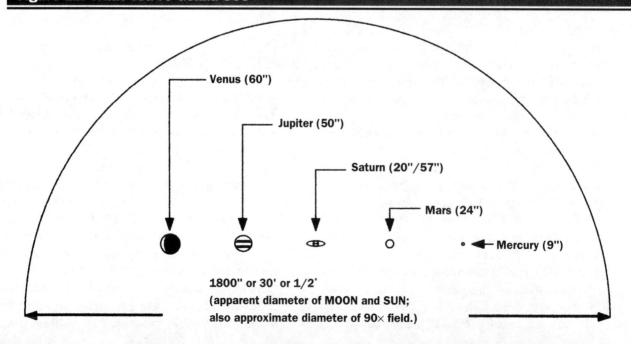

Venus (60")

Jupiter (50")

Saturn (20"/57")

Mars (24")

• ← Mercury (9")

1800" or 30' or 1/2°
(apparent diameter of MOON and SUN;
also approximate diameter of 90× field.)

bodies condensed out of local irregularities called "eddies." These few other bodies became the planets we know today. Their floor sweepings became comets and asteroids. What the broom couldn't even sweep onto the pan became—well, dust.

After all these objects became solid, they spent the next little while, about 500 million years, cooling down and learning how to drive around the Solar System. While they were learning, they ran into each other—often. The craters on the Moon and other planets offer ample evidence of that. That process left the planets in roughly circular orbits. It also resulted in the ejection of most of the comets into a spherical shell way out toward the edge of the Solar System. (You'll read in the next section just how far "way out" is.)

As to the location of all the "big stuff" in the Solar System, it's stayed pretty much the same for the last 4.5 billion years. Compared to the life of the Solar System, Earth's recorded history is the blink of an eye.

7.2

The Scale of the Solar System

I could just as easily have added this part of the general description of the Solar System to the previous section. I didn't because it's so much fun for us puny humans to try to comprehend the scale of something so many times bigger than we are.

7.2.1

But First, a Time Out to Look Back in Time

To define distance in this book, we're using inches, feet, yards, and miles when we can. However, the distance to the *nearest* star is 23,300,000,000,000 miles. That's a pain to write every time we want to refer to it, and besides, this number is for the *nearest* star. It only gets worse farther out.

In discussing the inner Solar System, we will deal with the distances by using millions of miles. However, scientists now believe that the suburbs of the Solar System extend about halfway to the nearest star, so, to avoid all those zeroes, we need to invoke that often-quoted-in-sci-fi-movies unit of distance, the *light-year*.

You'll remember from earlier sections that in a vacuum, light rolls along at 186,000 miles per second. Since the Sun is about 93,000,000 miles from Earth, it takes the Sun's light about 500 seconds to get here.

The speed of light is constant in the vacuum of space, so we can use how far it goes in a unit of time to create a unit of distance. The light-year is how far light can travel in one year.

But, let's back up a bit. We can also say that the Sun is about 500 light-seconds away, since it takes light that long to get here from the Sun. We could also say the Sun is 8.33 light-minutes away.

We can also define light-hours and light-days, but, of course, the light-year is the most common unit. Besides helping us become familiar with the light-year, thinking about the various units of distance based on how far light travels in a unit of time forces us to confront a fundamental fact about what we are really seeing when we look at celestial objects—the PAST.

When we look at the Moon, we do not see it as it is now. We see it as it was about 1.3 seconds ago, because that's how long it took the sunlight that bounced off the Moon's surface to reach our eyes. We do not see the Sun as it is now, either, but as it looked about 8.33 minutes ago. To see what it looks like now, we must wait 8.33 minutes. This little-thought experiment illustrates a simple but powerful point.

The farther away we look, the farther back in time we look.

So the only "now" is inside your head. Everything you see in the sky is in the past.

On a more practical note, the distance to the nearest star, 23,300,000,000,000 miles, becomes 3.98 light-years; at least we got rid of all those zeroes.

7.2.2

Now, REALLY, the Scale of the Solar System

First, the Sun is about 864,000 miles wide. The Earth is about 109 times the diameter of the Sun away from it, or about 93,000,000 miles from the Sun. The Earth is a bit less than 8000 miles wide. (The Moon is only 240,000 miles from Earth, so it is only 30 times the diameter of Earth away.)

The next significant distance is to Jupiter, the largest planet. That's about 484,000,000 miles from the Sun, or about 560 times the Sun's diameter from Jupiter. Once we go any farther than Jupiter, relating distances to the diameter of the Sun isn't relevant anymore.

The last item describing the scale of our Solar System is the distance to the cloud of comets astronomers believe lurks at the very edge of the Solar System. This is known as the Oort Cloud, after the Dutch astronomer who first proposed its

Figure 12: The Solar System

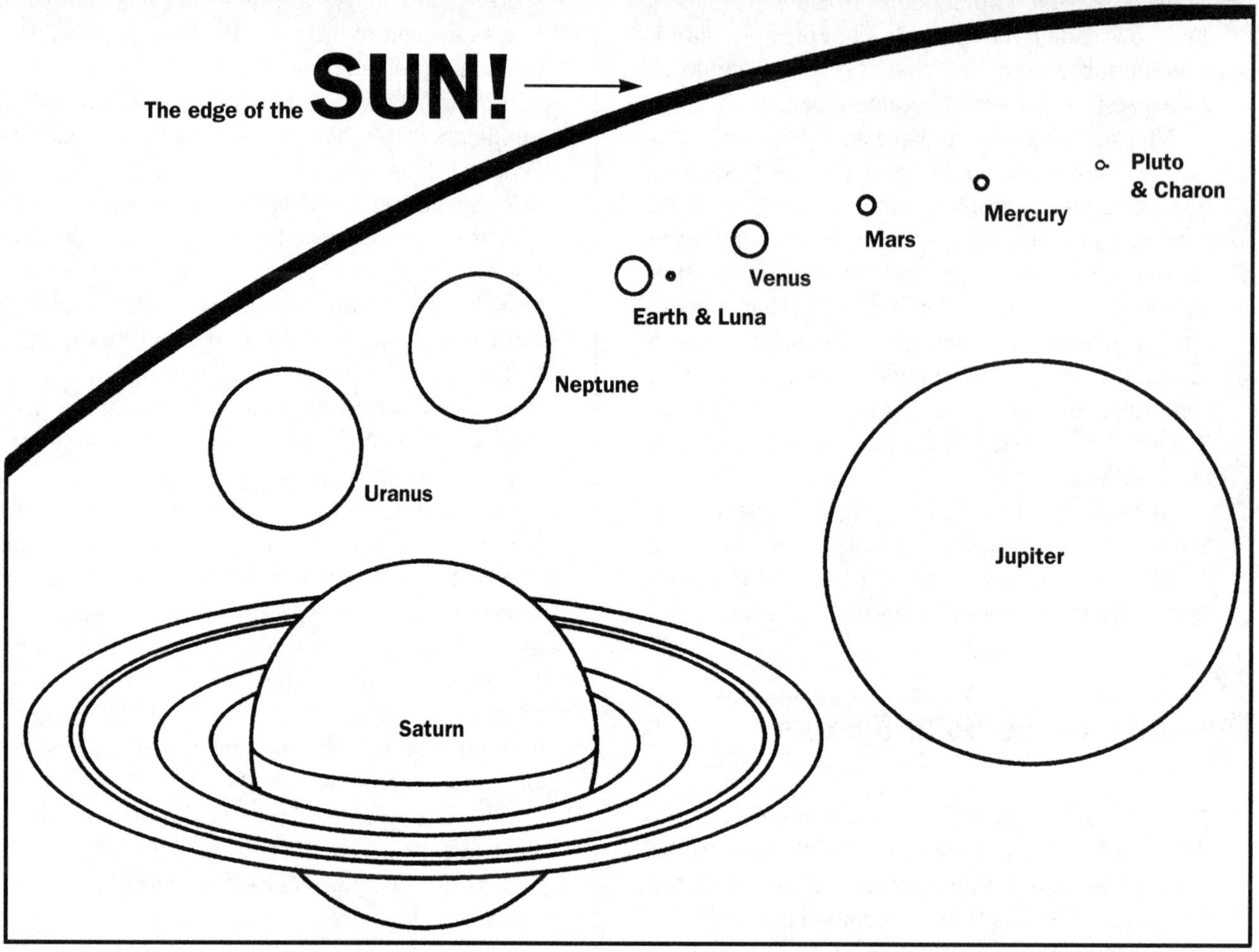

The edge of the **SUN!** →

Pluto & Charon

Mercury

Mars

Venus

Earth & Luna

Neptune

Uranus

Jupiter

Saturn

existence. Dr. Oort's theory requires the cloud to lie between 0.8 and 1.6 light-years from the Sun.

So, from the inner city, where we hang out, to the incredibly distant suburbs, where the comets hang out, it takes their light somewhere between 0.8 and 1.6 years to reach Earth. Our Solar System may be as much as 3.2 light-years across!

7.3

Old Reliable—the Sun

Besides pretty much dominating daytime, the Sun does supply literally all the energy we consume, and thus all the food we eat. One square foot of the Earth receives on the average the heat and light of two 60-watt light bulbs every second. This has been going on for five billion years and astronomers believe it will continue for about that long into the future.

Until physicists developed the modern theory of the atom, astronomers were at a loss to explain where the Sun was getting all that energy it spews out every second. Atomic theory provided the answer—the fusion of Hydrogen into Helium in the core of the Sun (and in every other star).

Since the Sun is the biggest and brightest object in the sky, it's relatively easy to observe, and it isn't observed nearly as often as it deserves to be. Possibly that's because it's easy to damage your eyes if you attempt to observe the Sun improperly. So pay attention to this section if you intend to observe the Sun.

7.3.1

O.K., So It Does Supply All Our Physical Needs; What IS the Sun, Anyway?

In one sentence, the Sun is an enormous, incredibly hot sphere of gas. Only it's not really a gas. It's beyond that. The Sun is so hot the electrons and atomic nuclei that would make up atoms cannot latch onto one another. They're all so energetic they just rush hither and yon, although I doubt that inside the Sun they could tell the difference between hither and yon.

What causes all the heat? The enormous pressure caused by all that matter attracting itself. That pressure causes the nuclei of Hydrogen to be driven together to form nuclei of Helium. Each resultant Helium nucleus masses a bit less than the two Hydrogen nuclei from which it was formed. That slight difference in mass is converted to energy according to Einstein's equation:

Energy = Mass × (velocity of light)^2 or $E = MC^2$

Even though the mass difference between the two nuclei is very small, it is overcome because the velocity of light is huge. Einstein further requires us to SQUARE it. So there's an incredible amount of energy being released every second down in the core of our Sun. The common term for this process is *nuclear fusion*.

It may have occurred to you that fusion of atomic nuclei to make a heavier nucleus is what happens inside a Hydrogen bomb. You're right. The difference is that it's occurring inside the Sun on a radically different scale. All the H-bombs in the world's arsenals would hardly be noticed if exploded within the core of the Sun. Just for fun, here's the amount of energy that's produced by the Sun every second:

380,000,000,000,000,000,000,000,000 watts!

In scientific notation, that's 3.8×10^{23} watts. All those zeroes are impressive, especially with that big exclamation point. But scientific notation is the lazy man's way.

A typical desk lamp puts out 50, or 5.0×10^1 watts every second.

Why doesn't the Sun just blow its top to get rid of all that energy? Because the Sun's own mass makes it tend to collapse because of its gravity. That gravitational collapse is offset by the energy of the nuclear fusion. This armed truce has been going on for about five billion years and will go on for another five billion years into the future.

Stars like our Sun stay pretty much the way the Sun is now for a very large portion of their lives. Other stars behave differently, as you'll read a little later. One thing they all have in common is that they all formed from a vast cloud of gas and dust.

This cloud of gas was mostly Hydrogen. There was also dust left over from the residue of the explosions of earlier stars. Since everything in the Universe attracts everything else due to gravity, a massive enough cloud will begin to contract as each of its particles attracts all the others. So "our" cloud also began to contract.

As the cloud contracted, it began to warm up as the particles bumped into each other. Since there was enough mass in the cloud, it contracted to the point where there was enough pressure at its center to start the fusion of Hydrogen nuclei into Helium nuclei, which released incredible amounts of energy. At that point, our cloud became a star.

Most clouds appear to have some rotation, and so did ours. As it contracted, its rotation increased for the same reason a figure skater's rotation increases as she draws her arms close to her body while she's spinning. The cloud flattened into a disc shape due to this rotation. The innermost portion of the cloud coalesced into the Sun. The leftovers at the edge of the cloud became the planets, asteroids, comets, and dust we find in our Solar System today. That all the large objects in our Solar System revolve around the Sun in the same direction as the Sun suggests that they must all have been formed out of the same original cloud.

While we're thinking about this, I'll tidy up a little. Here are a few notes about star formation and star evolution in general:

1. The cloud from which the Sun coalesced is usually called the *Solar Nebula*. For the simple discussion above, I found it clearer to write the word "cloud." You'll be looking at a lot of nebulas in the sky as you get farther along into astronomy. Many of them are where stars are forming today. When I tell you where to find them, I'll also identify which of these nebulas are "stellar nurseries."

2. The process of the creation of the Sun out of the Solar Nebula took about 30 million years, yet astronomers believe our Sun will be as it is now for at least 10 billion years.

3. The crucial factor in what a star is like is its mass.

   We've already looked at the time scale of the Sun's evolution, so let's consider what happens to a star that's much more massive than our Sun and see what happens to one that's a lot less massive:

   If the cloud from which a star forms is very massive, it contracts quickly, and "births" the star in only 100,000 years. This massive star will be much hotter than our Sun. It will also be much brighter. But it pays for all the glamour of living life in the "stellar fast lane." It consumes the Hydrogen fuel in its center at a prodigious rate and its life as a stable star lasts only 100 million years. This is a long time by the standard of a human lifetime, but very short when compared to the 10 billion years our Sun is expected to last. Indeed, we should be happy that our planet formed around such a plodding, average star. The Sun's long life has given us time to evolve.

If a cloud of gas is much less massive than "our Sun's nursery," it won't contract as fast, so it'll take 100 million years or so to develop enough pressure at its core to begin nuclear fusion and "light off." These stars are very long lived, but since they hadn't nearly as much Hydrogen in their cores to fuel nuclear fusion, they don't live as a stable star as long as our Sun will. Their lives last about two billion years.

4. I'll remind you of this in the next chapter, but it's so important I decided to include it here, too. The color of a star's light is what tells us its surface temperature. If a star is very bluish white, its temperature may be as high as 50,000° Kelvin, while a star that appears reddish may have a temperature of only 3000° K.

A star's color is the gross effect our eyes can see. But the spectrum of a star is the real "signature" that identifies it and reveals much about its nature. Astronomers use their analysis of extremely detailed spectra of stars to determine their chemical content. From its spectrum, they also infer other details of a star's nature, such as how fast it's rotating, if it has any close companion stars, and whether or not it's pulsating in size.

I deliberately broadened this discussion from its focus on the Sun to a discussion of stars in general. I did this because it seemed a good way to introduce to you the fundamental concepts that seem to govern what stars look like, and how they live their lives to become other objects in the sky that we can look at (and enjoy).

Instead of continuing this discussion to include the ways stars die and why their brightness often varies, I'm going to stop here. After all, the object of this book is not to bash you over the head with the weight of astrophysics, but to get you started enjoying the stars.

The rest of these important subjects I'll discuss as they apply to objects you can observe. The exception is that there's some more general information on stars in the next chapter.

7.3.2

What's the Structure of the Sun?

I've already mentioned that the fusion of Hydrogen and Helium nuclei is happening at the Sun's core. The temperature in the core is—(are you sitting down?)—15 million° on the Kelvin scale, or about 8.3 million° Fahrenheit. Regardless of the scale, that's hot.

Just above the core is a very deep layer of mostly Hydrogen waiting to be fed to the core. The energy being generated in the core radiates through this feed lot out to a cooler region just below the "surface" of the Sun.

This layer is called the zone of convection, because the energy bubbles to the "surface" much like steam on the surface of a pan of boiling water. The "surface" of the zone is the photosphere, which is what we see as the edge of the ball-shaped Sun when we happen to see the Sun through a thick cloud or through the atmosphere near the horizon at sunset. The temperature at the photosphere is *only* 6000° Kelvin.

In writing about the photosphere, I've put the word "surface" in quotes because there's really no hard surface on which a particularly well-insulated person could stand. It just looks like a surface when we observe the Sun.

Just above the photosphere, the Sun's atmosphere begins, consisting of a layer of glowing Hydrogen gas called the chromosphere, followed by a layer of very-super-ultra-heated plasma called the corona. The reason for the very-super-ultra adjective is this: The temperature of the corona is between one and two million degrees Kelvin. Only the fact that the plasma of the corona is extremely thin keeps it from putting out much light.

Beyond the corona is the solar wind, called that because it is an outward flow of charged particles that appears to continue at least past the orbit of Neptune. Even though the solar wind is incredibly thin, it really counts as the outer atmosphere of the Sun. That means we're orbiting the Sun in its outer atmosphere.

7.3.3

Observing the Sun—SAFELY

The Sun puts out gobs of energy at all wavelengths of light from radio clear out to X rays. Fortunately, our atmosphere protects us from most of the more energetic photons coming at us from the Sun. However, the atmosphere won't protect you from visible light from the Sun if you try to observe it improperly. So, we take a time-out to make sure you know how to observe the Sun safely, before going on to tell you what you can observe.

There are several ways to view the Sun safely. All have advantages and disadvantages. Before I tell you about them, please heed this warning:

There are many small imported telescopes lurking in basements, attics, and at garage sales. Many of them were imported when all their manufacturers included a small solar filter as part of their "kit." This filter was threaded to screw into the eyepieces of the telescope. When it worked, it blocked enough of the Sun's light to present a safe view of the Sun.

Filters like these work by absorbing the excess light, which means they get hot. If they can't get rid of the excess heat fast enough, they crack. The result? No more filter to protect your eye from the Sun's rays.

If you have such a filter, or know someone who has one, do not *ever* use it. Discard it immediately.

Now that I've given you the warning about the screw-in filters, here's how you can observe the Sun—safely.

7.3.3.1 Projecting an Image of the Sun

A low-power eyepiece can project an image of the Sun onto a card. Just cover your finder, then point the telescope at the Sun, adjusting its position until its shadow is a circle. You'll see a very bright spot on the card. By slowly moving the eyepiece away from the objective, you can focus the image. If the image is too small, move the card away from the eyepiece and refocus.

One advantage of this method is that several people can observe the Sun at the same time. If you use a clip board that can hold a piece of white paper, then project the image of the Sun on that, you can draw the positions of the various features with fairly high accuracy.

There is one serious potential disadvantage to using eyepiece projection: The Sun's heat may melt the cement that holds the lenses in an eyepiece together. To avoid this problem, limit the aperture of your telescope to about 2 inches, and don't project an image of the Sun for more than a few minutes at a time. (You can limit the aperture of your scope by cutting a piece of shirt-cardboard into a disc that will fit over the "sky-end" of your telescope. Then cut a 2-in. circular hole in the disc.) You could also try using an eyepiece that has no cemented lenses. However, they're hard to find these days.

7.3.3.2 Using a Filter Over the Objective

Most amateur astronomers observe the Sun by placing a reflective filter over the objective. Placing the filter over the objective means only a small portion of the Sun's light will even enter the telescope. Making the filter to reflect most of the Sun's light rather than absorbing it means the filter remains cool, and it won't add any distortion to your view of the Sun. Reflective filters are of two types— plastic-film and glass.

Plastic-film solar filters usually have two layers of extremely thin plastic on which the reflective coating is deposited in a vacuum. The film is so thin it doesn't blur the image of the Sun at all. It does, however, scatter a lot of light, so the image of the Sun through one of these filters looks a little washed-out compared to glass filters.

As you might imagine, a glass solar filter merely substitutes a thin piece of glass for the plastic. However, since the glass is many times thicker than the plastic film, it must be finished to the same degree of accuracy as any other optical component. That means an optical-glass reflective filter will cost considerably more than a plastic-film filter. However, it will provide a higher-contrast view of the Sun.

7.3.3.3 A Final Word on Safety

Just as you must when driving your car or using a power lawn mower, you really need to constantly think about safety when you're observing the Sun through your telescope. Always make sure that whatever filter you're using is securely fastened. Always remember either to cover your finder or place a filter over its objective whenever you're observing the Sun. Finally, never leave a telescope unattended so that a child could point it at the Sun.

Remember: Permanent eye damage or even blindness is the result of improperly observing the Sun with your telescope or binoculars.

7.3.4
What You'll See When Observing the Sun

Now that I've attempted to alert you again about the safety issue, what can you see on the Sun if you work up enough courage to look at it? The answer is—plenty!

7.3.4.1 Sunspots, Granules, and Faculae

No, that isn't the name of a multicultural law firm. It's the trio of features that adorn the photosphere of the Sun.

Sunspots are dark splotches on the Sun. They were first noticed by Galileo in 1610. (Since he didn't have a safe solar filter, Galileo must have observed the Sun through a thick cloud layer. Don't you do that!) Although they are very hot, sunspots appear dark to our view because they are considerably cooler than the rest of the photosphere.

The very dark center of a sunspot is called the *umbra*, while the dark gray area surrounding the umbra is the *penumbra*. The umbra has a temperature of about 4200° Kelvin, while the penumbra is much hotter at 5700°.

Sunspots are very social, which means they're often found in groups. They generally congregate in a band north and south of the Sun's equator, where they rotate with the Sun in a period of about 28 days. You can observe a large group of sunspots on the eastern limb, or edge, of the solar disc and watch it for a period of about two weeks until it disappears at the western limb. Most of the time sunspot groups do not last to reappear on the eastern limb again two weeks later, but some large ones have been seen to do that.

The average number of sunspots seems to vary in a cycle of about eleven years. The last maximum was around 1990, so the number of sunspots will decline for the next 5.5 years, only to reach maximum around 2001.

Sunspots are areas where the magnetic field of the Sun is extremely intense. They appear to result in some way from an interaction between the rotation of the Sun and the currents that create its magnetic field.

The vast bubbles of gas that carry energy to the top of the photosphere are called *granules*. Granules are several hundreds of miles in size, and last only a few minutes, certainly no more than ten. If you observe the Sun early in the morning when the seeing is steady, you should be able to see that the photosphere is covered with these granules. With exceptional seeing, watching a granule form and then dissipate can be exquisite.

Granulation of the photosphere is often interrupted by large bright splotches called *faculae*. These are usually easier to see near the limb of the Sun, but can be seen anywhere on the photosphere.

7.3.4.2   Observing the Sun from the Moon's Shadow

Possibly the single most exquisite sight in nature is the solar corona, the pearly white, many-stranded cloud that surrounds the Sun. The corona is extremely hot and extremely thin, so unless something completely blocks the light from the photosphere, we can't see the corona.

Fortunately, the Moon is almost the same angular size as the Sun. When it passes between the Earth and Sun, the Moon eclipses the entire photosphere, and reveals the corona in all its glory. It also reveals the chromosphere, that thin, red layer of Hydrogen gas between the photosphere and corona.

Specialized equipment allows the corona to be viewed without benefit of an eclipse. However, such equipment fails to provide the high-contrast view an eclipse does. For that reason, amateur astronomers seem to travel the globe from eclipse to eclipse. I've been to five since my first in 1963, and I've often heard fellow eclipse viewers discussing travel plans to the next eclipse within minutes of the end of the total phase of the present eclipse. It's that good.

7.4
Our Nearest Neighbor—the Moon

Luna, or the Moon, lies nearest to us in space. Since it is 2160 miles in diameter and only 240,000 miles away, its angular size is big—about a half a degree. Since it is so close and has no atmosphere, we can get a good view of the Moon even when the seeing is poor. It's also the easiest object to observe. Unless you have a big telescope, you'll not need a filter to cut down on the Moon's light.

7.4.1
The Nature of the Moon

There are several theories explaining the origin of the Moon. None of them explains perfectly all the observations of the Moon's nature. However, the theory with the fewest defects states the Moon was formed about 4.6 billion years ago when a planetary body, possibly the size of Mars, collided with the Earth, causing a large amount of material to be ejected into space. Some of the ejected material formed a disc orbiting Earth, out of which the Moon gradually formed. As the Moon gradually worked itself into a sphere, the more massive radioactive material tended to "sink" toward its center.

This collision almost totaled the Earth, and may have caused the Pacific Ocean to take up nearly all of one side of the Earth. Since all the various objects in the Solar System were still learning how to drive, the Earth and Moon continued to be bombarded with heavy debris until about three billion years ago, when the bombardment tapered off. Almost all the craters of the Moon were the result of this bombardment. The craters of Earth were also the result of this

Figure 13: The Moon

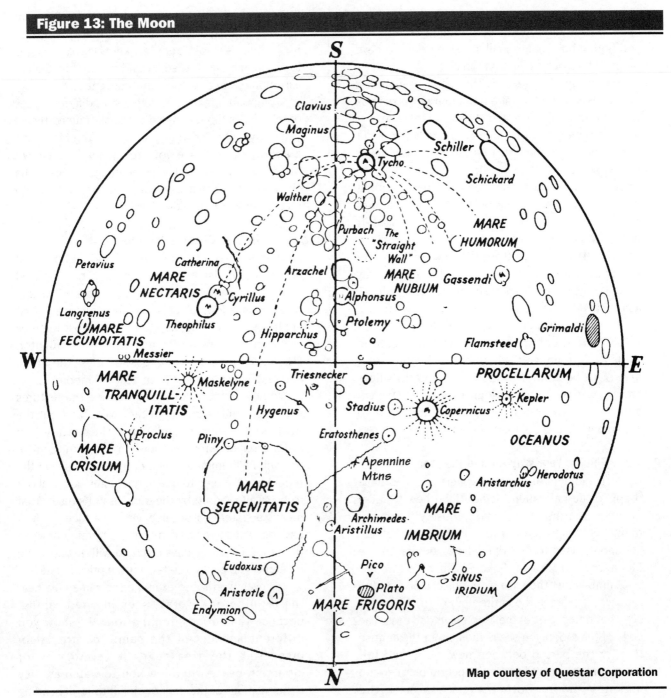

Map courtesy of Questar Corporation

bombardment, but they are not as obvious because our weather has eroded them.

Crater-covered highlands dominate the Moon, taking up about 85 percent of its surface. The other 15 percent is devoted to the "maria," which the ancients thought were seas, hence the use of the Latin word for "sea" to name them. The maria lie almost exclusively on the side of the Moon facing the Earth, so they give the impression that they cover a larger percentage of the Moon's area than they really do.

The maria are vast plains of solidified lava that flowed out of the interior of the Moon as it cooled.

The great heat of the Moon's interior was caused by radioactive materials decaying over the eons.

7.4.2

Observing the Moon

If you look at the full moon without optical aid, its two main features are obvious. There are dark, relatively smooth areas covering about 30 percent of the disc. The remaining 70 percent is brighter, appearing "busier" or rougher.

If you look at the Moon through binoculars, the craters and mountains pop into view. Most of the craters were caused by meteor impact, as you can demonstrate for yourself by finding a large crater and noting how many smaller craters seem to have plopped themselves right on top of it.

The mountains are also the result of meteoric impact. When you look at a mountain range such as the Apennines, note it follows a gentle arc. It appears the lunar mountains are literally the result of splashes of material that flowed away from major impact points and then froze in place!

A binocular view of the Moon also reveals bright lines, or rays, proceeding outward from several of the major craters. These are splash marks left behind when the last of the big craters were formed. The rays that radiate from the craters Tyco and Copernicus are the most obvious. The longest ray from Tyco can be clearly seen more than halfway across the Moon's disc.

A telescope reveals much more detail and two features that cannot be seen with binoculars. "Rilles" are long, thin, channel-like features that appear to be similar to features on Earth where areas have sunk from geological activity. "Domes" appear to be places where pressure built up so much that the crust of the Moon ballooned a bit.

7.4.2.1   Finding Your Way Around the Moon

Astronomers and cartographers seem to have dropped the ball when they decided how to label directions on the Moon. Before I justify this statement, let me give you a few definitions that are commonly used to describe where you are on the Moon (or the disc of any Solar-System object).

A limb is just the edge of an object's disc. At full moon, you can see the entire limb of the Moon. At all other times, part of the Moon is in darkness. The edge of the Moon you see is then either the sunrise line for the period between new moon and full moon, or the sunset line for the period between full and new moon. Both these lines mark the termination of the sunlit portion of the disc, so they are called the terminator.

Directions on the Moon are just as they are on Earth. If you were standing in Pocatello, Idaho, looking at the full moon as it crossed your meridian, the top limb is north, and the bottom limb is south. So far, so good.

Just as on Earth, east is to the right and west is to the left. That's O.K., so far as it goes. However, when we use this convention, we discover that the west limb of the Moon is "to the east" as far as direction on the Celestial Sphere is concerned and the

Moon's east limb is "to the west" on the Celestial Sphere. So long as you remember that, you'll have no problems. Most people I know have at times found themselves turned around due to this clash of astronomical and cartographic conventions.

Compounding the problem is that different types of telescopes flip the image of the sky differently. As you're trying to find your way around the Moon, it's a good idea to move your telescope in a known direction and see which way the image moves in the eyepiece. Doing that is usually better than trying to remember which direction is which.

7.4.2.2   When to Observe the Moon

If you've been to the top of a mountain or hill and observed distant objects, you may have noticed you can see best just after sunrise or before sunset, when the shadows objects cast are the longest. That applies to observing the Moon, too. Most features are seen best when they are near the terminator. (The exceptions are the gross features, the maria and rays, which are best seen near full moon.)

You'll see the most if you observe the features near the terminator, where they will stand out the best. Remember, though, that the feature you observe near the terminator when it is at the sunrise line will be illuminated differently when it's near the terminator fourteen to fifteen days later as it is about to be swallowed up by the darkness of sunset. That means a crater you observe when the Moon is four days old will again be seen near the terminator when the Moon is twenty days old, and will appear quite different. You get two craters for the price of one.

It's even worth it to examine the same area near the terminator several times in one evening, because the varied illumination will show you different aspects of the same feature. Also, inspecting the area near the terminator for mountain peaks will allow you to watch as they either rise out of the dark or sink into it, depending on whether you're observing early or late in the lunar month.

To plan when to observe what feature on the Moon, consult the lunar map. Features near the east limb of the Moon will be observed best within the first five days after new moon, and then for the first five days after full moon. Features near the center third of the Moon will be seen best around first quarter, and then half a lunar month later around third quarter. Features near the west limb of the Moon are best observed just before full moon, and just before new moon.

7.4.2.3 How to Observe the Moon

If you wish to view the Moon through binoculars, you'll see more if you support them to reduce vibration. One way is to mount them on a tripod, using the mounting socket found in the front of the hinge. If you can't mount them on a tripod, hold your binoculars in both hands and lean your elbows on the roof of a car or van.

When viewing anything at high power through a telescope, you want to be sure you are comfortable, so your eye will be relaxed. That means you should be seated while viewing. If you can, orient the eyepiece of your telescope so you don't have to crane your neck to see through it.

You may find that a neutral density filter will reduce the glare from the Moon and allow you a better view. Many of the companies that advertise in the various magazines sell filters of this type.

Finally, since there is usually plenty of light, you can view the Moon at as high a power as the seeing will allow. You'll probably find that using a power that yields an exit pupil smaller than 0.5 mm will not show you any more than one yielding a pupil between 0.5 and 1.0 mm. That's the exit pupil at which the human eye resolves best.

7.5

Observing the Inner Planets

Mercury, Venus, the Earth, and Mars are usually described as the inner planets. They also fall into the category of "rocky worlds" since they all have rocky surfaces and relatively thin atmospheres compared to the outer planets.

7.5.1
Mercury

As you read in Chapter 3, Mercury never strays very far from the Sun, so it is very hard to observe. Until Mercury was visited by the Mariner 10 spacecraft in the early seventies, we knew little about it. It is a world very similar to the Moon, covered with craters. It also has large, relatively smooth areas that resemble the lunar maria, suggesting that lava flowed out of its interior early in its history. Since there is no atmosphere to distribute heat, the sunlit side of Mercury is exceedingly hot—700° Kelvin. The night side of Mercury would be just as hostile, with a temperature of only 90° Kelvin, corresponding to minus 300° Fahrenheit.

There's little to observe on Mercury except its phases. It appears yellow when viewed through a telescope because its rough surface tends to scatter the short wavelengths more than the longer ones. Removing more blue from the Sun's light as it strikes Mercury leaves a yellow-white hue.

7.5.2
Venus

Astronomers often refer to Venus as "Earth's twin," because the two planets are so similar in size. That, however, is where any similarity ends.

Venus is a world of extreme heat, heavy clouds, and extremely little to view but its phases. Not only can you see nothing except the tops of Venus's clouds, you wouldn't want to even consider visiting.

A visit is out of the question because the clouds that shroud Venus are 96 percent carbon dioxide gas with a layer of sulphuric acid high in the atmosphere. (If you could harvest this stuff, you might corner the market on drain cleaner.) The atmospheric pressure is 90 times that on Earth. Since the atmosphere traps a lot of heat from the Sun, the temperature on the surface of Venus is 750° Kelvin. Russian spacecraft have landed on Venus and photographed its rocky, desert-like surface. They didn't last long.

So, there's nothing to see on Venus except its pearly white cloud tops. Because you can't see through these clouds, the only thing left is to observe Venus's phases. Oh yeah, the way to prove that "the bright light in the sky after sunset" is not a UFO is to show your neighbors the phases of that favorite UFO—Venus.

7.5.3
Mars

Mars has fascinated planetary observers since humans began observing it through telescopes. That's because it's the only planet we can easily observe that has both a visible surface and only enough weather to make things interesting, but not enough to obstruct our view most of the time. Although Mars is not an easy object in a small telescope, you can see a few major details through one—easily enough to impress your friends and neighbors.

7.5.3.1 The Nature of Mars

Mars is little more than half Earth's diameter. Since it is small and not very massive, its gravity can't hold onto as much atmosphere as Earth can. Much of it has leaked off into space. Its thin atmosphere meant that Mars had little protection from meteors, so it is pockmarked with craters. But some craters on Mars appear to be of volcanic origin. The largest, Olympus Mons, is a much larger mountain than any on Earth. It is 360 miles across and rises 15 miles above the surrounding plain. Scientist have not ruled out the possibility that some Martian volcanoes may be active, although none have been observed.

It appears there was once flowing water on the surface of ancient Mars, because there are formations resembling river valleys, complete with creeks and streams feeding the main rivers. However, this flowing water either existed before the last craters were formed or there was not enough flowing water to erode them all. The largest of these river valleys, the exquisite Valles Marinaris, is 2500 miles long!

The water that once flowed freely on Mars is now locked in the polar caps that are the easiest features on Mars to see in small telescopes. Planetary geologists also believe there is a lot of water stored in permafrost under the surface of Mars. Some of the moisture frozen in the polar caps does melt and become part of the Martian atmosphere, because the polar caps shrink in the Martian summer and expand in winter.

The Martian atmosphere is mostly Carbon Dioxide (96 percent), with a smattering of Nitrogen, Argon, Oxygen, and water vapor. Mars's atmosphere is so thin its atmospheric pressure is less than 1 percent of the pressure we feel here on Earth. Despite its thinness, Mars's atmosphere is able to generate extreme winds, sometimes as high as 80 mph. These winds move a lot of dust around the planet. Sometimes, the entire disc of Mars looks yellow, as a planet-wide dust storm rages, rather than the reddish brown we are used to seeing.

The two Viking spacecraft that landed on Mars in 1976 carried experiments designed to scoop up Martian soil and to perform experiments on the samples in a search for life on the Red Planet. They found no evidence of the kind of life scientists believed might be present. Although that suggests Mars is a lifeless world, it does not prove no life of any type exists there. Nor does it prove life never existed on Mars. A mission staffed by humans may be required to make a variety of tests before we pronounce Mars lifeless. Such a mission would be very costly, so it would appear it would have to be a worldwide effort, not exclusively an American one.

Mars has two small satellites, Phobos and Deimos, discovered in 1877 by an astronomer named Asaph Hall while he was trying out a new 18-inch telescope his observatory had just installed. (The telescope passed the test!) Close inspection of Phobos and Deimos by various deep-space probes shows both are nonspherical, pockmarked with craters and grooves, suggesting they are small asteroids captured by Mars sometime in the past.

7.5.3.2   Observing Mars

Despite its relative closeness to Earth, Mars is not easy to observe. As discussed in Chapter 3, Mars makes a really close approach to Earth only every fifteen to seventeen years. At those oppositions, it can attain an angular diameter of 24". The other oppositions, occurring at 26-month intervals, feature a maximum Martian disc of 12–16". So, if you have a 60-mm refractor, which resolves to about 2", you will be limited to seeing details between one-sixth and one-eighth the diameter of the disc. Only the polar caps and one major feature, Syrtis Major, are readily seen in a 60-mm telescope.

Larger telescopes reveal much more detail, including, very occasionally, the so-called canals. Although many very experienced observers have seen them, the spacecraft that have photographed the planet have been unable to detect them. It appears they are the result of the human eye's tendency to see a straight line when presented with an image of many very small objects that are approximately in a row.

The spacecraft images revealed the craters, volcanic mountains, and the river valleys, all features that were not regularly seen or photographed from Earth.

Humans see and photograph one set of features on Mars from Earth, then we send spacecraft there. The spacecraft images seem to show a radically different world. What gives?

The answer is we see and photograph much more gross features from Earth than the image features recorded by the spacecraft. Remember, the seeing scale in Chapter 5 rated as "excellent" seeing a night when the "seeing disc" is no greater than 0.5" in diameter. That means we can only discern those features that size or larger. If the disc of Mars were 20" in diameter, and the seeing limited the resolution to 0.5", we could only resolve high-contrast objects that were 0.5"/20" as large as the disc. That corresponds to 2.5 percent of the diameter of the disc. Since Mars is about 4200 miles in diameter, that means we can only see high-contrast objects that are 105 miles in diameter.

Since there are many objects that do not show high contrast with respect to the objects around them, the 105-mile figure is quite optimistic.

For example, many observers saw the giant volcano Olympus Mons as a darker spot on Mars's surface. They called it Nix Olympica. Sometimes they saw the white clouds that formed when water vapor in the atmosphere condensed as it was forced to a high altitude by flowing up the surface of the volcano. But, they never saw Olympus Mons as a volcano, because they couldn't resolve it.

Although there are a few reliable reports of a Martian crater or two being observed from the Earth, they're not seen by the average observer. What, then, do we see on Mars?

The answer is mostly, we see piles of dust. The dark features we see are areas where the Martian winds have scoured the surface free of dust. The "red" areas are where dust has congregated. When we see a feature on Mars change, it's because the wind has either removed some dust or dumped some there.

The dark areas are reddish-gray. However, our eye sees them as green against the red of the lighter-colored areas. The fact they were perceived as green and seemed to grow in the Martian summer led some astronomers early in this century to believe they were areas of vegetation, possibly under cultivation. They believed that the canals were elaborate irrigation ditches. Alas, at least to that extent, Mars *is* dead.

If you observe Mars with a modest telescope, you can see a number of dark areas, the polar caps, and the occasional yellowish dust storm. If you observe Mars regularly, though, you'll see more than that as you train your eye-brain combination to see more and more subtle details. Mars is not an object whose details jump out at you. But, if you put in your time, observing Mars can be very rewarding.

One technique for improving your view of Mars is to use a red filter to enhance the contrast between the dark and bright areas. Such a filter makes the whole planet appear darker than without the filter, but it darkens the reddish-gray areas more than the lighter ones, so you see the whole planet with increased contrast.

You can also improve your view of Mars by remembering that as the magnification goes up the image contrast goes down. To maximize contrast when observing Mars, raise the power until you see no improvement in the image, then drop the power one notch.

Observing the Outer Planets

The Outer Planets consist of the four so-called "Gas Giants," Jupiter, Saturn, Uranus, and Neptune, and an enigma near the edge of the Solar System, Pluto. The Gas Giants are called that because their predominant aspect is that they are huge compared to the inner, rocky planets and their thick atmospheres are their dominant features, not their surfaces.

You'll likely find observing Uranus and Neptune to be of little interest. However, Jupiter and Saturn continually vie for recognition as the *pièce de résistance* of Solar System objects. I never tire of observing either of these magnificent planets.

The Nature of the Gas Giants

As a group, the Gas Giants are quite similar, even though they don't look very much alike. Each has a thick atmosphere rich in Hydrogen and Helium. The pressure of the atmosphere steadily increases until the Hydrogen becomes liquid. Jupiter and Saturn are massive enough that near their small solid cores the Hydrogen may be forced into a liquid metallic form. Uranus and Neptune each have a similar gaseous atmosphere, incredibly thick layers of liquid, and, possibly, a small solid core.

Although they are very similar, each of the four Gas Giants looks different. That's because of several factors:

1. As you go out from the Sun, each is smaller and less massive than the last. The more massive a Gas Giant is, the more pressure near its core. The more pressure, the higher the core temperature. Also, as you go out from the Sun, each receives less heat from the Sun than the last. The colder the atmosphere, the less active it is. So, Jupiter is a riot of belts and zones, while Saturn is less volatile, and Uranus and Neptune, while interesting scientifically, have very pale belts and zones, which cannot be glimpsed in any but the absolute best telescopes, under the best seeing conditions.

2. Although each has a predominantly Hydrogen and Helium atmosphere, each atmosphere has different impurities that cause the cloud tops of each to have different colors.

Each Gas Giant rotates quite fast, with Jupiter taking the shortest period—9 hours, 55 minutes. The others take longer, but all four are rotating fast enough that they're each wider across the equator than from pole to pole. Jupiter is the only one of the four that is noticeably squashed. Its equatorial diameter is about 6 percent greater than its polar diameter.

Of course, it's the cloud system of a Gas Giant that we see rotating. What makes these systems so interesting is that the various belts in a Gas Giant's atmosphere rotate faster than the zones between them. At the boundary, considerable turbulence is generated, which can lead to the chaotic view we get when the seeing is good.

As you might imagine, there are storms of considerable magnitude in atmospheres as deep and thick as those that make up a Gas Giant. Jupiter's Great Red Spot is an enormous, oval-shaped whirlpool whose long axis is *twice* the size of Earth. Astronomers have observed the GRS for at least 350 years. The GRS is thought to be the result of material spewing from lower regions of Jupiter's atmosphere. It is among the highest and coldest of all the features found in Jupiter's cloud tops.

White features are thought to be a bit lower and warmer than red features in a Gas Giant's atmosphere. Jupiter always seems to feature white spots that are all considerably smaller than the GRS. Occasionally, a white spot or two can be seen circling the edge of the GRS. However, it takes several days to see this motion.

White spots are often found rolling along in Saturn's belts also. Occasionally, Saturn treats us to a lollapalooza of a spot. These seem to reappear every thirty years. Since the last one was in 1990, it looks like we'll have to wait until 2020 to see the next one.

Brown spots, which some people call barges, are lower and warmer than the white spots. As with other spot-like features, they tend to congregate at the border between belts and zones, which would indicate they are the result of turbulence between the two.

On Jupiter and Saturn, bluish features are seen less often than features of other colors. They are lower and warmer than any of the others. Of course, due to a difference in the impurities in the atmospheres of Uranus and Neptune, blue-green or blue are the dominant colors.

Besides spots, the turbulence at the border between belts and zones often results in part of a belt being dragged into a zone. These features are called festoons.

Satellites: Since they are all very massive, each of the Gas Giants has a large collection of satellites, some of which may not have formed with its Gas Giant. Rather, some of the satellites may be cap-

tured asteroids. We know of sixteen moons of Jupiter, while Saturn has seventeen, Uranus, fourteen, and Neptune, eight moons.

Although it's not something we can observe from Earth, it's worth noting that the satellites of the Gas Giants are all worlds with their own radically different features. Io, one of the satellites of Jupiter discovered by Galileo, features active volcanoes. Others, while not quite so demonstrative, are equally interesting.

Rings: Depending how you define a moon, each of the Gas Giants can be said to have thousands. They're the individual moonlets that make up the rings surrounding all four Giants. Saturn, of course, has a magnificent system of rings that is one of the most exquisite sights in the heavens. Unfortunately, the rings around Jupiter, Uranus, and Neptune are only detectable by spacecraft imaging or through great observatory telescopes here on Earth.

7.6.2
Observing the Gas Giants

7.6.2.1 Jupiter

When you look at Jupiter in even a small telescope, you will immediately see that its flattened disc is girdled by dark belts running east and west across it. Between the belts are the lighter-colored zones, although they could just as easily be described as light-colored belts.

If you observe Jupiter with a larger telescope, say, a 6-in. reflector, detail between the belts becomes distinct, and the Great Red Spot might be visible. White spots and dark barges become visible. Often a festoon will dip down from one of the belts into a zone.

The Great Red Spot has, at times, been almost a fire-engine red. At others, it is salmon-colored and almost indistinguishable from the other features. When it is at its palest, it can often be recognized because it displaces the south equatorial belt, creating the Red Spot hollow. So, even if the GRS can't easily be seen, it still makes its presence felt.

Spots are not the only phenomena of the Jovian disc that seem to change from time to time. The south equatorial belt has faded nearly to invisibility on several occasions, most recently in 1991. The more you look, the more you'll see on Jupiter's ever-changing cloud tops.

I've often found I can increase the contrast between the various features on Jupiter by using a light blue filter. However, if you have access to filters of other colors, try them, too. Since Jupiter has features of many colors, other filters can improve your view of other features.

7.6.2.2 Jupiter's Galilean Satellites

All the details I've just mentioned will only be visible if the seeing is good. If it's not, you can still watch the four Galilean satellites as they move around the planet. Don't expect to see them actually move from second to second. However, you can see them move relative to each other and Jupiter itself after ten to fifteen minutes.

As you observe the Galilean moons, you'll soon notice that some seem to move fairly fast, while others appear to be positively glacial. The mini-table shown below will give you the details:

| Jupiter's Moons | | | | |
Name	Number	Diameter (miles)	Distance (miles)	Period (days)
Io	I	1128	262,000	1.8
Europa	II	975	417,000	3.6
Ganymede	III	1635	665,000	7.2
Callisto	IV	1491	1,170,000	16.7

I've included the Roman numeral assigned to each of the Galilean satellites because that's the way they are identified on charts published in the magazines and almanacs to show their relative positions. The charts are usually printed vertically, and graph the positions as wavy lines, called sine waves. The time scale is usually printed down the left side of each chart. (Remember that these charts are usually done in terms of UTC.) Jupiter itself is drawn as a pair of straight lines running down the middle of the chart.

To determine where the satellites will be, just find the time you wish to observe them and lay a ruler across the chart at the point corresponding to that time. Where the ruler intercepts each line is where you'll see that satellite. Since these charts are drawn to scale, you can make a good estimate of how far each moon will be from Jupiter itself.

If you look really closely at one of these charts, you'll notice the lines pass first in front of the "Jupiter bar," then behind it. That indicates the moon will pass in front of Jupiter's disc (a transit) or behind it (an occultation).

A transit occurs when a moon is moving toward Jupiter's disc from the east. If you see a moon approaching the disc from the west, it will pass

behind Jupiter. Before opposition, when Jupiter is in the morning sky, a moon's shadow will precede it across the disc. After opposition, a moon's shadow will follow it across the disc. A moon's shadow will be a black dot on the cloud tops, obvious even in a small telescope. Since the Galilean moons are similar in color to Jupiter itself, they are seldom visible as they transit. I've seen the disc of Io against the disc of Jupiter only a few times. On all these occasions, the seeing was exquisite. So is the telescope I use for observing the planets.

When an occultation occurs, the satellite won't disappear suddenly. It will take a while to fade out, because its disc can't go behind Jupiter all at once. Also, Jupiter's atmosphere will refract light around the cloud tops a bit, lengthening the disappearance.

Occasionally, one moon will eclipse another. You can find out when that is predicted to occur by consulting any of the annual almanacs or one of the magazines.

If you have no telescope, don't fail to observe the Galilean moons with binoculars. To steady your view, hold the binoculars with both hands and rest your elbows on a fence or the roof of a car or van. If Jupiter is very high in the sky, lie on a reclining lawn chair or on a blanket on your lawn, then rest the eyecups of your binoculars on the bridge of your nose to steady them.

7.6.2.3 Saturn

Saturn puzzled Galileo when he first observed it through his telescope. Since he couldn't quite resolve the rings, his impression of Saturn was of a large disc with two smaller ones on each side of it. It remained to Cassini to first see the rings for what they are.

Saturn's rings are among the most exquisite sights in the heavens. I never tire of seeing them. However, there's more to Saturn than its rings.

Saturn's cloud tops are similar in layout to those of Jupiter. Saturn is colder than Jupiter, so the turbulence in its cloud tops is not as obvious. However, with the exception of the Great Red Spot, Saturn's cloud tops all have the same features: belts and zones, white spots, dark spots, and an occasional festoon. Remember that Saturn's disc is only a maximum of 19.5" in diameter, which, coupled with the subtlety of its features, makes it much harder to observe. Making up for that is the butterscotch-color of the belts of Saturn.

Using colored filters to enhance various features is even more important when observing Saturn than Jupiter. The most important filter is the light blue you can also use with Jupiter. This filter increases contrast between the belts and zones just as it does with Jupiter.

7.6.2.4   Saturn's Magnificent Rings

I cannot remember ever showing someone the rings of Saturn for the first time and having that person NOT be entranced. Even when the seeing is bad, the rings are beautiful. When the seeing is good, they're exquisite.

Saturn's rings extend about 2.26 times the diameter of the planet itself, so they turn a planet never more than 19.5" in diameter into an object that is as much as 44" across! But there's a catch to this. Twice in each trip Saturn takes around the Sun, the Earth passes through the plane of the rings. Since the orbital period of Saturn is 29.5 years, about every 15 years we see the rings edge-on. When that happens, we see the rings of Saturn as the "line of Saturn." This will next happen in 1995. After that, the rings will appear to widen until 2002, when they will begin to flatten again.

Observers had occasionally glimpsed considerable detail in the rings of Saturn, and planetary scientists had theorized they were made up of the debris of a satellite that came too close to Saturn and was torn apart by tidal forces. It wasn't until the Voyager spacecraft made close approaches to Saturn in the mid-eighties that their true nature was confirmed. The rings are made up of billions of small satellites of Saturn, each in a separate orbit. Their size ranges from dust to gravel to the occasional Volkswagen-sized particle. Most of the particles are probably chunks of ice, ice-covered rock, or a mixture of the two.

As you might imagine, a system with so many particles doesn't stay quite the same for long, just as a crowd of people leaving a stadium is not static. However, we humans do tend to order ourselves when we're in groups, simply by staying to the right and trying to avoid collisions. The rings of Saturn also exhibit quite a bit of order. There are many divisions in the rings, and the small ones have been glimpsed from Earth under the best of conditions.

There are two major divisions in the rings, which separate the whole system into the three major rings. The outermost, the A ring, is separated from the brightest ring, the B ring, by Cassini's division. This division was discovered by the man who discovered the rings themselves. Since Cassini's division is approximately 2" wide when we view it from Earth, it can just be seen with a 60-mm

telescope when the seeing is steady. If you can't see it, you'll not see much else, either.

Encke's division separates the A ring into an outer dark and an inner brighter ring. It can be seen under excellent conditions at each end of the rings, but not all the way around the A ring. Another feature of the rings that can only be glimpsed under excellent conditions is the C ring, which is called the "crepe ring" because it is very tenuous, resembling translucent crepe paper. If the seeing is very steady, you can see the crepe ring partially obscuring the ball of Saturn itself.

At opposition, the shadow of the ball of the planet cannot be seen on the back of the rings. However, before or after opposition, you should be able to see the shadow of the rings of Saturn. Once you gain some experience observing Saturn, you may be able to notice that Cassini's division is a very dark gray compared to the shadow. Once you can notice the difference, you've become a very experienced observer, indeed.

A *Final Note*: Some scientists believe that the rings of Saturn are not a long-term feature of Saturn, but may have formed only recently. If that is the case, we are indeed lucky to be living in a time when the rings of Saturn grace our Solar System.

7.6.2.5   Observing Saturn's Satellites

Like its more massive cousin Jupiter, Saturn has managed to collect an amazing array of satellites, led by the largest satellite in the Solar System, Titan. Besides this satellite, Saturn has sixteen other moons.

Titan is easily seen through binoculars or a small telescope. You can recognize it as (usually) the brightest star near Saturn. If you observe Saturn for several nights, you'll see Titan move with respect to the planet. A large telescope will allow Titan's disc to be resolved. Its reddish color also becomes apparent in a large instrument. At its maximum excursion from Saturn, Titan can be found about three times the diameter of the rings out from the edge of the rings.

Several more satellites are visible in moderate-sized telescopes. You can hunt for them by deliberately placing the planet itself just outside your field of view. This will allow you to see the fainter ones unhindered by the glare of Saturn.

7.6.2.6   Uranus

Uranus was the first planet to be "discovered," as all the brighter planets have been known since we humans have been observing the sky. Sir William Herschel found it in 1781 with a telescope just a bit over 6 inches in aperture. It is unique in that its axis of rotation is tilted so much it rolls along its orbit sideways. Of course, like any rotating body, Uranus's axis always points in the same direction. During the nineties we'll be looking at Uranus's South Pole.

About all there is to see when you observe Uranus is a tiny blue-green disc a bit less than 4" in diameter. Since Uranus is much farther out than Saturn, it's colder, making its belts and zones invisible to all but a handful of Earthbound observers. Since it is brighter than sixth magnitude, you can find it without the aid of a telescope on a clear dark night.

Uranus has an amazing collection of satellites, as well as its own ring system, both of which are invisible with small telescopes.

7.6.2.7 Neptune

In its last encounter with a planet, the Voyager 2 spacecraft found a wonderful blue-green world with a Great Dark Spot and exquisite white clouds. Unfortunately, Neptune is as boring to observe from Earth as it was wonderful to see through Voyager's "eyes."

Why? Neptune is so far from us it's only a bit more than 2" in diameter. That means Neptune's disc appears no wider than Cassini's division in Saturn's rings! It usually takes a 6-inch telescope, good seeing, and more than 200× even to resolve Neptune into a bluish disc.

Neptune has an incomplete system of rings, and we know of eight moons. None of these can be seen in the average amateur's telescope.

7.6.3

Pluto (and Charon)—the Solar System's Frozen Enigmas

Pluto was discovered in 1930 by one of the most beloved of all amateur astronomers, Clyde Tombaugh. Although he was then an amateur astronomer, Tombaugh was working as an observer at Lowell Observatory in Flagstaff, Arizona.

His work consisted of taking exposures of the ecliptic area of the sky with a wide-field camera. He would then take an exposure of the identical area a few weeks later, then compare the two identical photographs in a device called a "blink comparator," which allowed him to look back and forth at the two images very rapidly. The stars wouldn't appear to move, because the scale of the photographs was too small. Only an object in our Solar System would appear to "jump" back and forth because of its motion around the Sun. Pluto, of course, "jumped" as Clyde Tombaugh "blinked" its discovery photographs.

Because it is so remote and so small, we know little of Pluto. It is a twin world, as it and its only known moon, Charon, orbit each other every 6.39 days. (Charon was discovered in 1978 by the very sharp-eyed James Christy, an astronomer at the U.S. Naval Observatory, who noticed that many images of Pluto were slightly elongated.)

Pluto is probably a chunk of debris left over from the formation of the Solar System. Both the surfaces of Pluto and Charon are covered with ices of water and other organic compounds, possibly Methane and Ammonia. It's likely that Pluto and Charon are tidally locked to each other, and present the same face to each other, as the Moon does to Earth.

Pluto's orbit is the most eccentric, or noncircular of all the planets. Since it is near its closest point to the Sun—perihelion—Pluto is now closer to the Sun than Neptune and will remain so until the year 2000.

No one can observe Pluto through a telescope. All you can do is detect it because, at magnitude 14, Pluto is at the edge of visibility of all but the larger amateur telescopes. I've detected Pluto twice, but it is unlikely I'll ever generate enough enthusiasm to try again.

You can use a modified version of "blinking" to find Pluto. Both of the major magazines publishes a finder map for Pluto at the beginning of the year, usually based on a photograph of the sky through which Pluto is passing. Locate the field that contains Pluto for the night you wish to detect it. Find that field in your telescope. This will require an 8-inch telescope and very steady seeing.

Make a sketch of the field, using different-sized dots to represent different star brightnesses. If all goes well, one of the dimmest stars is Pluto. But which one? To find out, repeat the process the next clear night. You guessed it: The star that moved is Pluto.

7.7

The Leftovers—Comets, Asteroids, and Meteors

It might be hard to believe, but comets, meteors, and asteroids have a lot in common. They're what was left over when the planets coalesced from the disc of gas and dust that was part of the original Solar Nebula. Much of the time, these "leftovers" are hardly noticeable, but sometimes they make themselves known in spectacular fashion.

7.7.1
Comets

For much of our history, humans viewed comets with dread. We thought they brought famine, pestilence, or other disasters. The pendulum has now swung full circle: The appearance of Halley's comet in 1986 had all the trappings of a major Hollywood event, even though there have been many comets that have appeared in my lifetime that were more spectacular than Halley. A great comet can be a spectacle. So what are these occasional visitors to our sky?

7.7.1.1   Comets as Dirty Snowballs

The leading theory of the nature of comets is that they are quite similar to what you'd get if you took some snow and some dust from your vacuum cleaner and made a "dirty snowball." The one difference between your homemade snowball and a comet, besides the size, is that the comet's snow is not just frozen water. It's also Methane, Ammonia, Carbon Dioxide, and a host of other organic chemicals. This theory was first proposed by comet expert Fred Whipple in 1950. The European and Soviet probes to Comet Halley in 1986 verified Whipple's theory.

Most astronomers believe comets are much as they were when the Solar System formed, so studying them can give us clues to what things were like then. The catch is comets are quite small, and we seldom get to observe them except when they come close to the Sun. The snowball itself is called the nucleus and is seldom more than a mile or two in diameter. It's also irregularly shaped. (Halley's nucleus looked like a potato.)

Here's what happens as a comet approaches the Sun, along with the major features we can observe:

As the outer surface of the nucleus is melted by the Sun's heat, the ice is quickly converted directly to gas, without ever existing as a liquid. This frees the dust that was trapped in the ice. All this material surrounds the nucleus in an opaque cloud resembling ground fog. In about half of all comets observed, this fog-enshrouded nucleus can be seen as a tiny star-like feature.

Surrounding the nucleus is a less dense but much larger cloud called the coma. It may be 50,000–60,000 miles in diameter and reflects so much sunlight it obscures the nucleus. For a short time, when it's near the Sun, a pipsqueak-sized object of a mile or so in diameter can pass itself off as an object nearly as big as Jupiter.

The great mass of the comet's frozen nucleus continues in its orbit as the outer surface is melting to form the coma. The gas and dust freed by the melting don't have a lot of mass, so when they reach the outer part of the coma, they aren't moving in the same orbit as the nucleus.

Ultraviolet light from the Sun tears the gas apart into charged particles. When we break up atoms into charged particles, we call the particles ions. These ions are carried along by the solar wind, and flow straight back from the coma to form the gas tail or ion tail of the comet. On long-exposure photographs, the ion tail appears blue because the small size of the ions won't allow them to reflect much of the long-wavelength red light from the Sun. We see only the shorter-wavelength blue light.

The dust grains freed in warming have little mass, colliding with other grains of dust lurking in the path of the comet. This dust tail is often curved from interaction with the local dust. Since the dust grains are much bigger than the ions, they reflect the red end of the solar spectrum better than the blue, so we see the dust tail as distinctly reddish on long-exposure photographs.

The blue and red colors for the ion tail and the dust tail are almost never seen visually, as our eyes just don't discern colors at the extreme end of the spectrum unless the light levels are fairly high. However, if you see a very straight comet tail, it's almost certain to be the ion tail. If it's curved, it's dust.

The size of a comet's tail can be truly astonishing. It's not uncommon for a tail to be 10 million to 20 million miles long. A few have been as much as 100 million miles long, approximately the distance of the Earth from the Sun. It's no wonder our forebears feared the advent of a great comet.

(If you are a fan of the two syndicated shows "Star Trek: The Next Generation," and "Star Trek: Deep Space Nine," you may have noticed the computer graphics of various celestial objects that are part of the lead-in to the show as the cast names are flashed on the screen. The next time you watch, notice that the beginning of each episode of "TNG" takes the viewer through the two tails of a comet. The beginning of "DS9" features a comet cruising by, complete with gas and dust being thrown off the nucleus.)

7.7.1.2 The Orbits of Comets

Just like all other objects, comets move about the Sun in elliptical orbits. Since they are not massive objects, comets are often "captured" by the gravity of the Gas Giants, especially Jupiter and Saturn, each of which has families of comets whose orbits they have modified. All these comets have orbital periods that have simple mathematical relationships to the period of the capturing planet.

Since a comet can lose as much as 1 percent of its mass to melting every time it approaches the Sun, they can't be very long lived. That means there must be a fresh supply of comets to replace those that melt themselves into comet oblivion. This argument led the Dutch astronomer Jan Oort to conclude that a cloud of comets must exist somewhere. That many comets seem to have extremely long periods led Oort to believe the cloud was located very far away from the Sun.

Although there is no direct evidence for its existence, most astronomers accept the existence of the Oort Cloud. Its incredible supply of comets was possibly ejected from the neighborhood of the Sun after most of the planets coalesced from the Solar Nebula, so they lurk in the nether regions of the Solar System waiting for something to happen.

Very occasionally, a passing star causes some comets to fall inward toward the Sun. These objects are of the most interest to astronomers, because of the likelihood that they've never been in the vicinity of the Sun since they were ejected into the Oort Cloud. That means they may be exactly as they were just after the Solar System formed.

Some comets that approach the Sun from the Oort Cloud have orbital periods exceeding 30,000 years. So, even if they're not on their first approach to the Sun, they've not been near its heat for a substantial portion of their history and they may be relatively pristine.

It appears all the comets that have short periods originated in the Oort Cloud, plunged into the vicinity of the Sun, and were captured by one of the Gas Giants, converting their orbits into ones with periods of a few years or tens of years. Some of these comets remained part of a Gas Giant's family, while others were attracted by close approaches to other objects, which shifted them into other orbits.

Astronomers can observe short-period comets, also called periodic comets, several times in a row. Not even Methuselah would have been long-lived enough to see a comet with a 30,000-year period return to the Sun again and again.

7.7.1.3 The Brightness of Comets

Comets do not generate their own light. When you see one, you see the Sun's light reflected off the various layers of gas and dust that enshroud the "snowball." That means the brightness of a comet is dependent on only three factors:

1. The first is how close the comet comes to the Sun. The closer it gets, the more energy it receives, and the more material will melt off its surface. The more gas and dust melts off, the brighter the coma and tail of the comet will be.
2. The second factor that controls how bright a comet will be is whether or not it has been in the vicinity of the Sun before or very often. If it hasn't been by the Sun in a while, it may have swept up a lot of dust in its travels. This dust would block much of the Sun's heat from reaching the ice it wants to melt. As the saying goes, "no melting ice, no comet bright."
3. The first two factors control the intrinsic brightness, the techno-term for how bright a comet would be when seen at a standard distance. The final factor affecting how bright a comet will appear to us on Earth is how far away it is. Remember that an object 50 million miles away will appear only a quarter as bright as one that's 25 million miles out. If you move it to a distance of 100 million miles, it will be only one-ninth, or about 11 percent, as bright as the same object seen at a distance of 25 million miles.

To have a truly bright comet, it needs to make a close approach to the Sun, much of the dust it's swept up in its travels must be burned off, and it must be close to us. Usually, two of the three is good enough to produce a "nice" comet, while one of the three produces a relatively dim one.

7.7.1.4 Observing Comets

Since many comets are big, dim objects, your first view should be with the lowest power your telescope can provide, which will maximize the contrast and allow you to see the tail if the comet has one. (Some comets don't get close enough to the Sun to allow enough material to be melted off to form a visible tail.)

For particularly large comets, binoculars often provide the best view of a comet's tail. Remember to use averted vision to view the dimmest portions of the tail. Also, by sweeping your binoculars' field across the tail rapidly, you'll see more of it.

Observing the coma and nucleus requires higher power. As usual, keep bumping up the power until doing so results in no improvement in the view. At high power with steady seeing, you may notice irregularities in the coma near the nucleus. These may appear to be jets of material flowing off the nucleus. These are jets of gas blasting off the surface of the comet.

You can't expect to see these jets actually moving. However, if you take a break of as little as an hour, you could notice the shape of the area around the nucleus change. That could be a change in the flow of material or it could be evidence of the rotation of the nucleus.

Comets can appear anywhere in the sky. However, they will usually be at their brightest when they are near the Sun, which means that you often have only a few minutes after sunset to find some comets just after they have passed close to the Sun. Careful planning will pay off in such situations. Another reason to be ready early is that leisurely watching a comet appear out of the gloom as darkness falls can be an experience you'll never forget.

Some amateurs spend many hours searching for comets. Part of their motivation is the joy of discovery. However, since a new comet is named for the discoverer, the attendant notoriety is also a powerful motivation. If you wish to try your hand at this, consult a book on comets.

7.7.2
Meteors

Almost everyone has seen a "shooting star." Others call them "falling stars." Although most people learned in school that they're not stars at all, they still call them that. So what? If someone is out under the night sky with you and sees a meteor and calls it one of those other things, don't "get all het up" about it. Use it as an opportunity to show your friend how much you know about meteors.

7.7.2.1   Meteoroids, Meteors, and Meteorites

The dust and gravel released from a comet as its outer layer melts is the prime source of meteoroids. These small particles litter space, and are the source of many dramatic moments in science-fiction novels.

Every day the Earth runs into many of these particles. But, before the particle can reach the surface, its high velocity creates friction with Earth's upper atmosphere and the particle burns up. The meteoroid that was minding its own business until it got run over by a bus is, for a brief moment as it's burning up, a meteor. If it reaches the ground, a meteor is called a meteorite.

Very few meteoroids are big enough to survive being a meteor and impact the Earth to become a meteorite. Those that do are valuable to scientists because they provide clues to the chemical makeup of the early Solar System. However, literally many tons of the ash that remains after meteors have burned up in the atmosphere gently float to Earth every day to mix with everything else.

For no apparent reason, large meteorites are usually named for the post office nearest where they fall. However, if you find yourself flattened by a meteorite, astronomy might make an exception and name it after you. Don't hold your breath about this. There are no known examples of an individual being killed by a meteorite.

Often, a meteor is sufficiently big to have a fiery trail following it. The techno-term for such a fireball is bolide. It's nice to know that term, but most astronomers I know call them fireballs.

If you see a fireball break up into several pieces and then slow down, it may mean it will reach the Earth's surface. Such a fireball may also be accompanied by sound effects, most of which are sonic booms, as the object is still moving faster than sound. Other fireballs leave smoke trails. If you see a fireball, be sure to look at its trail with binoculars. They often last for several minutes, and it's fun to watch the winds in the upper atmosphere make them twist and turn.

On any given night, you can expect to see five to six meteors an hour. However, on many nights that number can dramatically increase. These events are called meteor showers. Here's why they happen:

At some point in its life, enough of the outer layer of the nucleus of a comet may melt away that the comet can no longer hold itself together. It breaks up into many small "dirty snowballs." Since the surface area of these minicomets is much greater than the larger body that preceded them, the ices melt much faster and flow into the gas tail of the comet. That leaves a concentration of dust and gravel where the nucleus of the comet used to be.

Gradually, the dust and gravel spreads out around the orbit of the former comet. If the Earth intersects that orbit, it runs into some of the debris left over from the comet. On Earth we see this as a meteor shower. Even comets that haven't broken up yet are the sources of meteor showers.

Shower meteors seem to originate from one point in the sky, called the radiant. They're not really radiating from that point in the sky. That's just the effect of our perspective. The name of each shower is associated with the constellation within which the radiant lies.

Here's a list of the major meteor showers:

Major Meteor Showers

Shower	Date	Hourly Rate
Quadrantids	Jan 3	30
Eta Aquarids	Apr 23	10
Delta Aquarids	Jul 30	15
Perseids	Aug 12	40-60
Orionids	Oct 21	15
Leonids	Nov 16	6*
Geminids	Dec 13	50
Ursids	Dec 22	12

*The Leonid Shower is seldom of much note. However, every 33–34 years, the Earth intersects the orbit of the spent comet when the debris left over from the comet's nucleus comes around the orbit. The last time this happened, 1966, areas of the southwestern U.S. received intermittent meteor showers with rates (verified by photography) of 150,000 meteors per hour. I saw that shower through a small hole in the clouds from northern Illinois. It looked as if stars were falling through the hole! The next "super Leonid shower" will be in 1999 and 2000.

7.7.2.2 Observing Meteors

Meteors are best observed from a place with a clear, dark sky. The best time is after midnight, when the Earth's rotation is moving the nighttime side of the Earth toward the direction it is moving in its orbit. In this geometry, we're "slammin' them meteoroids" with all we've got. Before midnight, we're trying to catch up to the dusty little buggers.

The specialized observing equipment needed for observing meteors consists of a lawn chair, bug spray, sleeping bag, and your two eyes. Have a pair of binoculars ready to examine any smoke trails. (Don't get too comfortable. I did this once and fell asleep. I awoke at sunrise with a coating of ice on my moustache! True story.) If you're looking for shower meteors, face the direction of their radiant.

7.7.3
Asteroids

Asteroids are big meteoroids. In fact, there may be somewhat large chunks of rock lurking inside the nuclei of some comets. When all the ice has melted, the biggest chunks just keep on in the same orbit. Other asteroids, especially the really big ones, are not likely to have been associated with a comet.

Most asteroids orbit the Sun in a belt between the orbits of Mars and Jupiter. It was an early belief that this so-called asteroid belt was debris left when a planet broke up. If that's the case, it wasn't a very large planet, as the largest asteroid, 1 Ceres, is a bit more than 600 miles in diameter. By the way, the number 1 preceding the name of the asteroid means it was the first one discovered.

Unlike comets, asteroids are named *by* their discoverer, not *for* her or him. Some astronomers whose research regularly leads to the discovery of asteroids get a real kick out of naming them after a friend, loved one, or their hometowns.

There is a class of asteroids whose orbits cross Earth's orbit. Many scientists believe that one of these collided with our planet about 65 million years ago. The ensuing dust cloud enshrouded the Earth and killed much of the vegetation, which depended on sunlight. It is thought that the dinosaurs, who had no food, became extinct as a result.

In 1908, a smallish asteroid (or largish meteoroid) impacted Earth in the Tunguska region of Siberia, destroying a forest within a 25-mile radius and killing a few reindeer. Had it hit a major populated area, it would have done as much damage as a fair-sized H-bomb.

Although I find the research into the nature of asteroids interesting, I find them a bore to observe, like Pluto. You can say you've seen one, but beyond that, there's not much to see. The largest and brightest asteroids can become as bright as sixth magnitude, so they're not hard to identify. If you want to identify dimmer asteroids, use the same method I outlined for finding Pluto. Both the magazines publish finder charts for an asteroid or two each month.

CHAPTER 8

Outside Our Solar System— the Milky Way Galaxy

If you look at a dark sky with just your eyes, there are only three objects you can possibly see that are not part of our home galaxy, the Milky Way. Those three objects are galaxies themselves.

If you view the sky with your binoculars or a telescope, the only class of objects you can view that are not in the Milky Way galaxy are other galaxies. Fortunately, the Milky Way is filled with wonderful objects to view. This chapter tells you a bit about each class of object, as well as how to view them. It also treats viewing the Milky Way galaxy as a whole.

Oh, yes. Those three objects that aren't in our galaxy? They're the two satellite galaxies of Earth, called the Magellanic Clouds, and the Great Andromeda galaxy. Only the Andromeda galaxy can be see from mid-northern latitudes. You must travel to Hawaii or farther south to see the satellites of our galaxy.

8.1

About the Milky Way in General

8.1.1
What IS the Milky Way?

A previous guide for beings hitchhiking around the Milky Way referred to how big space is. Truer words were never written.

The Milky Way galaxy is an enormous, disc-shaped cloud of stars, gas, and dust about 100,000 light-years across. Our Sun shares the galaxy with about 100 billion other stars. Not even that many stars makes things crowded. The average separation of stars is about five light-years. The whole mass is rotating around the center, making one trip around it in something like 200 million years.

If you could look at the galaxy edge-on, you would see it has midriff bulge. That's because there is a concentration of stars at its very center. The disc itself is about 2000 light-years thick, while the bulge is five times that thick. Running along the center of the disc is a band of dark matter, its dust lane. More about that in a bit.

Looking at the galaxy from far above one of its poles paints a very different picture. Now it's a vast whirlpool of bright spiral arms with dark dust lanes between the arms. As you probably surmised, the bright arms are filled with stars.

At the end of each dust lane, the dust has spilled out and coated the middle of the edge of the galaxy all the way around it. So, when we look at our galaxy edge on, it looks like a flat, inverted Oreo™ cookie (light on the outside, dark in the middle) with a midriff bulge!

We humans always seem to want to know where we are, whether it's in the AP poll of college basketball teams, or in the Milky Way galaxy. So, I'll tell you: We're about two-thirds of the way out from the edge, and a little below the plane of the galaxy. More specifically, we're located in a spiral arm known as the Orion arm after the constellation that uses it as its hunting ground. The arm inward from us is called the Sagittarius arm, after the constellation that lies at its heart. Just outward from us is the Perseus arm.

8.1.2
Observing the Milky Way Galaxy

You'll seldom see the Milky Way from urban areas. In fact, I doubt if more than a few city dwellers have ever seen it. As an example, I point out that an astronomy student who was visiting my club's observatory in the mountains northwest of Los Angeles first saw the Milky Way, then remarked to me: "Gee, it's too bad about the funny-shaped cloud over there. I'll bet it's going to block out a lot of our views!"

The most useful item of equipment for observing the Milky Way is the reclining lawn chair. This piece of gear allows you to relax and enjoy the majesty of our galaxy in comfort. It also allows you to scan up and down this river of light with a pair of binoculars. The ones I use often are a pair of 7×50 surplus naval binoculars, but any pair will give you a great view. So what do we see when we view the Milky Way?

Our language only allows us to say that we're looking at the Milky Way galaxy. However, it's important to remember that it's all around us. What we discern as the Milky Way in the sky is what we see when we look in the plane of our galaxy's disc. It's there that we see the cloud-like river of light that stretches right across the sky from horizon to horizon.

In late summer here in the northern hemisphere, the galaxy runs from southwest to northeast. When we look at the summer Milky Way, we're looking inward toward the center. Because we're near the inner edge of the Orion arm, the dust lane between it and the Sagittarius arm obscures much of that arm. What we see are vast dark rifts in the Milky Way, as if there are islands in the milky river. Inspecting the dark rifts with binoculars can often reveal quite a bit of detail.

Since our Solar System is located a bit below the plane of our galaxy, we look under these dark rifts to see into the Sagittarius arm. The center of our galaxy is located (far) inward from the Sagittarius arm, so when you're looking toward Sagittarius, you're looking inward toward the galactic center. Since there are more spiral arms inward from us than there are looking outward, the summer Milky Way is brighter than the winter Milky Way.

When you look at the winter Milky Way, you're looking through our home arm, the Orion arm, and away from the galactic center. There are many more bright stars, but fewer stars on the whole, so the winter Milky Way is not so bright as the summer version. However, the cold winter sky is much drier than the summer sky, so less light is scattered by water vapor, allowing us to see the winter Milky Way with higher contrast than we might in a warm, humid climate.

When we look at any part of the Milky Way in a telescope, we find that the "milkiness" is caused by the combined light of stars too dim for our unaided eyes to resolve. So, the first type of Milky Way object we'll discuss will be the stars.

8.2

Stars

After the Moon and planets, the stars are the most obvious things we see in the sky. And yet, they are so remote from us we can never observe their surfaces, except, of course, for the Sun, from which we've learned the most about stars. We're limited to observing how bright stars are, including changes in their brightness, their colors, and how they behave if there are multiple stars in a system. Only extremely advanced systems allow astronomers to make images of the surfaces of some of the larger nearby stars.

8.2.1

A Bit More on the Nature of Stars

I decided to tell you most of what you'll likely need to know about the nature of typical stars in Section 7.3 when I discussed the nature of the Sun. In this section I'll present a few other facts about stars, along with a few reminders of some important things to know about stars that explain why they look the way they do.

1. How bright a star appears to us as we observe it from Earth depends on two main factors: how far away it is, and how intrinsically bright it is.

   You probably already have a feel for how the brightness of an object changes with its distance. So, if we have a star of certain brightness and we move it twice as far away from us, it will only appear a fourth as bright. If we move it three times as far away, it will only appear a ninth as bright, and so on.

   The catch is that all stars are not alike. They differ in brightness, color, size, age, and in the material from which they formed. Right now, we're only interested in their brightness.

   To establish a level playing field when comparing the brightness of stars, astronomers moved them all to a standard distance, about 33 light-years. Then they measured their brightness relative to each other. The brightness of each star at this standard distance is called its intrinsic brightness. Of course, this feat of moving all the stars to a standard distance wasn't done in reality, but on paper.

   Example: The brightest star in the sky is Sirius, in the constellation Canis Major. (See the ALL-Sky map for February.) We see Sirius as magnitude −1.4. Its distance is about 8.4 light-years. However, if we moved Sirius to the standard distance, it would only appear as a +1.4-magnitude star. That's almost a three-magnitude difference. It's still a bright star. But it's the brightest star in our sky only because it's so close.

   Our Sun appears as an object of magnitude −26.7. If it were at the standard distance of 33 light-years, it would only appear to be a star of the fifth magnitude, just bright enough to be seen with the naked eye.

2. After you've spent a little time observing the sky, you'll begin to notice that some stars appear a

bit reddish, some more yellowish, some seem to be pure white, and some shine with an almost blue-white light. Your eyes are not deceiving you. The colors are real. The reason it takes a while for you to notice the colors is that it takes a while for your eye-brain combination to learn to separate these slight differences in color.

Astronomers have been able to measure the temperatures of stars. They've found that the redder a star is, the cooler it is, and the bluer it is, the hotter. So, a red star might have a temperature of as little as 3000° Kelvin, while the hottest blue-white stars have a temperature of as much as 50,000°K.

3. When you think about the sizes of the various types of stars, it's a very complex discussion. Some very old stars can be very cool, and thus very red, but are also huge. Because their surface area is also huge, they are very bright in total, even though any one small area on such a star puts out a rather feeble amount of light.

But, all red stars are not giants, and all giants are not red. There are giant stars that are much hotter and bluer than any others. There are also red stars that are considerably smaller than our Sun. It takes several chapters of a college astronomy text to sort it all out.

8.2.2
Multiple Star Systems

Possibly a third of all stars are actually multiple stars, that is, systems in which two or more stars revolve around a common center. Some multiple stars cannot be seen through a telescope, but their existence can be inferred by the way they affect the spectrum of the "main" star. Other multiple stars can be seen through a small telescope. Some multiple stars can actually be detected moving around one another. However, it takes almost a lifetime and very accurate equipment to make those measurements. As you might suspect, the bulk of multiple stars are double stars, although systems with as many as seven components are known.

Many amateur astronomers like to observe double stars because they can be a challenging test of a telescope's optics and of the observer's skills. Double stars whose components are of the same or similar brightness are the easiest to "split." As you observe doubles that have bigger and bigger differences in brightness, you'll find them harder and harder to split. When splitting double stars, use the highest power that seeing conditions will allow.

There's another reason to observe double stars: The colors of the components are often different and can present a beautiful contrast. To observe a double star for its color contrast, use only as high a power as you need to just split the star. That maximizes the color contrast.

8.2.3
Variable Stars

Many stars vary in brightness. Astronomers study these variable stars for clues to their nature. Amateur astronomers have been contributing to the observation of variable stars for years by sending their observations to:

American Association of Variable Star Observers
25 Birch Street
Cambridge, MA 02138

There are many types of variable stars. Some are doubles whose components eclipse each other, causing us to see a periodic change in brightness. This is the only type of variable star caused by conditions external to the star.

Other variables are stars that pulsate in size (and thus brightness) in response to as yet poorly understood conditions in their interiors. Some of these variables seem to follow very predictable relationships between their intrinsic brightness and their period of variability. Others seem to vary their brightness aimlessly. Still others vary over periods as long as one to two years.

Some variable stars seem not to follow any pattern at all, but seem to explode or flare in brightness. Going over every one of these is beyond the scope of this book. However, more and more amateur astronomers are observing these types of variable stars. The AAVSO is always looking for more. If these phenomena strike your fancy, I urge you to contact AAVSO.

The brightness of some stars varies in the extreme, and only once. These are stars making catastrophic, one-time changes in their conditions.

The least violent of these is the *nova*. This event occurs when one of the stars in a binary system has contracted to a very dense state because it's run out of nuclear fuel to keep it inflated. It sucks some material from the outer layers of its companion, which triggers a brief but intense flare of energy. A nova will brighten by so many magnitudes it seems to appear out of nowhere. (That's how it got its name. Nova means "new" in Latin.) It will then gradually fade until it is as dim as it was before the event.

The most violent of suddenly variable stars is the *supernova*. In this event, an old massive star collapses suddenly (due to its own gravity) because there is no more nuclear fuel to support fusion near its center. It can't keep itself inflated any longer. The shock wave of the initial collapse bounces off itself when it reaches the center of the star, creating a titanic explosion. The word titanic is insufficient to describe a supernova. The typical supernova will put out more light for a few weeks after the explosion than the entire galaxy in which it resides! Only a tiny remnant of the original star is left after a supernova explosion.

Since supernovas occur in our galaxy only a few times each century, you'll not often get a chance to observe one. Novas happen more often, so you might get a chance to observe some during your astronomical career. And, estimates of the brightness of novas are useful to science. However, all other types of variable stars are observable night after night. It's *those* you can observe to make the biggest contribution to scientific research.

8.2.3.1   Observing Variables

You can observe the changes in brightness of the brighter variable stars with binoculars, although most observers use a telescope. The technique is simple:

You find the variable star and compare it to a star of known brightness that's close to the expected brightness of the variable. If the variable is a little brighter than the comparison star, you must find a brighter comparison star. Once you've done that, you know the variable is between the brightnesses of the two comparison stars.

Over the years, skilled observers have found they can estimate the brightness of variable stars to within 0.1 magnitude. So, if you have two comparison stars of 7.7 and 7.9, and you're sure the variable is between them, you can assign a magnitude of 7.8 to the variable.

8.3
Some Catchall Notes on Deep-Sky Objects

Amateur astronomers have divided what you can see in the sky into three major categories: (1) the Sun, Moon, planets, and other Solar System objects; (2) the stars; and (3) the so-called "deep-sky objects," such as nebulas, clusters, and galaxies.

This division has little to do with any rational organization of the Universe. Rather, it reflects what can be observed from where. Most objects in our Solar System can be observed from anywhere in a large city. As you get a bit away from the city, you can observe many stars. (This is mostly confined to splitting double stars.) Only when you get far away from the city can you observe the "deep-sky objects," which are presumably so-named because they are deep in the darkest skies. (A better name might be "dark-sky objects.")

I chose to organize this book by starting with what is close to us, the Earth, then proceeding through the Solar System to the Milky Way galaxy, then to the external galaxies. However, you should know that all the objects I describe from this point on are known as "deep-sky objects" by amateur astronomers.

8.3.1
Finding Deep-Sky Objects

Since they're "deep in the dark sky," these objects may be too dim to see in your finder telescope. What to do? Most amateurs star hop. All that means is that you point your telescope at a star near the dim object you wish to view, then move it to another star a bit nearer. Finally, you move your telescope so the point relative to the last visible star where your map shows the object to be. To aid you in star hopping, I've included on each ZOOM map a circle the size of the field of the typical finder or binocular.

8.3.2
About the Magnitude of a Deep-Sky Object

Professional astronomers are very interested in the brightness of individual stars because they use that information to help work out the physics of stars.

However, they're less interested in the brightness of deep-sky objects, as that has little to do with their scientific studies. The magnitudes of deep-sky objects are estimates that have been made by amateur observers and can be inaccurate by as much as several magnitudes.

One factor that affects this inaccuracy is that deep-sky objects are spread out over the sky, so the magnitude assigned to one of them is the equivalent to what we would see if it were concentrated at one point, like a star. This means a nebula with a magnitude of nine would appear much dimmer than a star of that magnitude, because its light is spread out over a lot of sky.

8.4

Nebulas

As humans began surveying the sky with telescopes, they observed many types of cloudy objects. The Latin word for cloud, nebula, became the word astronomers used to describe those objects. All nebulas consist of gas or dust or both. The various types of nebula you can see in the sky differ either in how they are illuminated, or in how they formed.

To make your understanding of nebulas less cloudy, each of the following items tells you about the various types of nebula, and how to observe each type.

8.4.1

Emission Nebulas

These clouds consist mostly of gas that is absorbing ultraviolet light from nearby hot, young stars. Although the gas absorbs the UV from the stars, it immediately reradiates it in the visual region of the spectrum, mostly in the red end, but some in the green. Since the gas that makes up the nebula is emitting light, it's called an emission nebula.

Many emission nebulas are star nurseries, where infant stars are forming. The hot, young stars that provide the power to the nebula are just the first wave of a baby-star boom. Since nebulas spawn stars, it's not surprising that most emission nebulas have clusters of stars associated with them.

If you photograph an emission nebula, you find their predominant color is red. However, when you look at an emission nebula, you see either a white cloud or one that's tinged with green. The red is there, but your eyes are not very sensitive to red, so you see little of it. Your eye is most sensitive to green, so that's what you see. Once you become a serious observer, you may begin to see tinges of red and purple in some of the brighter nebulas. That's because your eye-brain combination learns to detect the very subtle shades of those colors to some extent.

8.4.1.1   Observing Emission Nebulas

Your first view of an emission nebula should be low power so you can see it at its highest contrast. As you boost the power, the contrast goes down, but you begin to see delicate swirls in the cloud. Reserve the highest power for bright, highly detailed regions of the nebula. Finally, remember that dim

objects like nebulas brighten when you look at them using averted vision.

In the last ten years, special nebular filters have become available. These color filters are designed to block light in regions of the spectrum in which nebulas emit little light, and to pass those regions in which the nebulas emit most of their light. These filters can dramatically increase the contrast and allow you to see much more of a nebula than you could without them. In many cases, these filters will allow you to see nebulas that you couldn't see without them. The disadvantage is that the stars no longer look white, but take on a green hue.

8.4.2

Reflection Nebulas

Some nebulas seem only to shine by the reflected light of nearby stars. When we look at them, we see only a dim white cloudy object, with no greenish tinges at all. If they are photographed, they appear mostly bluish. That's because the dust grains in the cloud reflect blue light from the star better than the red light at the other end of the spectrum. Much more of the red passes right through the cloud, so it's not reflected so much to our eyes.

8.4.2.1   Observing Reflection Nebulas

The technique for observing these nebulas is the same as for observing emission nebulas. However, you'll find that a nebular filter works poorly in observing reflection nebulas. That's because the nebula is reflecting starlight across a large part of the visual spectrum, not in the narrow regions typical of emission nebulas.

8.4.3

Dark Nebulas

Dark nebulas result when a cloud of gas and dust is neither excited into emission by a nearby star nor is any light being reflected from it. We see a dark patch with stars all around it.

The various dust lanes in our Milky Way galaxy are examples of huge dark nebulas, but they can be seen up and down the Milky Way. Often, there are dark nebulas between us and either an emission or a reflection nebula. These often make an interesting contrast.

8.4.3.1   Observing Dark Nebulas

Binoculars are excellent for observing the large dark nebulas in our galaxy. A telescope at low power also gives excellent views. Space doesn't permit me to include many descriptions of dark nebulas in the commentary that accompanies each month's maps. That may be a bonus, because you can discover them for yourself, as I did. I've found sweeping the Milky Way for dark nebulas to be among the most rewarding hours I've spent observing the sky.

8.4.4
Planetary Nebulas

As they age, some stars shed their outer atmospheres into large shell-shaped nebulas, which we see as small, circular patches of light. William Herschel, who discovered this class of objects, called them planetary nebulas, because they resembled the faint disc of a planet in his telescope. Their only connection with planets is that they surround a star, just as planets must.

After the star creates the planetary nebula, it has been compressed a bit, which heats it up. It emits quite a bit of UV light, which excites the gases in the nebula in the same way as in an emission nebula. A planetary nebula is a shell-shaped emission nebula surrounding a star.

Planetary nebulas happen often, but don't last long, so they're not everywhere you look. Also, they are seldom very big, since as they expand they fade from view.

8.4.4.1   Observing Planetary Nebulas

With only a few nearby exceptions, planetary nebulas are small, so high power is necessary to see any detail in them. Only a few are even marginally impressive in small telescopes. However, if you have the opportunity to view one of the bright planetaries in a large telescope on a dark night, don't pass it up.

8.5
Clusters of Stars

Since many stars form out of a single great cloud of gas and dust, clusters of stars can be seen all over the sky. Some clusters formed out of rather loose clouds, while some formed out of very intense clouds. How they formed is what separates the two types of cluster.

8.5.1
Galactic Clusters

The most common type of cluster found in the sky is the galactic cluster. Since the stars in this type of cluster are spread out and often have no recognizable center, they're also called "open clusters." They formed out of the same cloud of gas and dust, so they often have nebulosity associated with them. Astronomers have also found that they all seem to be traveling through space in roughly the same direction.

Galactic clusters are confined to the "lens," or disc of our galaxy, so you'll find many more of them in the Milky Way than other parts of the sky. However, since every star you can see is in our galaxy, galactic clusters can be found anywhere in the sky.

8.5.1.1   Observing Galactic Clusters

Scanning the Milky Way galaxy with a pair of binoculars will net you many galactic clusters. In general, a low power will show a galactic cluster best. One of the fun aspects of viewing open clusters is comparing the colors of the stars within them. Often, a few red stars will provide a wonderful contrast with the white stars in the cluster.

8.5.2
Globular Clusters

Globular clusters are the titans of our galaxy. They are enormous symmetrical clusters of stars arranged in a spherical shell or halo around the periphery of our galaxy. The globular club is very exclusive: There are only about 150 in the galactic halo. They're also very old, having been formed as soon as the galaxy was old enough to form stars.

The typical globular cluster has about 100,000 stars in it. They're really packed together near the center and their numbers decrease as you move toward the edge. The close ones are among the most striking sights in the sky, while the distant ones can be very challenging objects, since they might be as much as 80,000 light-years away.

8.5.2.1   Observing Globular Clusters

Only the brightest globulars are interesting in a small telescope. At medium power, the typical globular will show a nebulous blur at its center, with a sprinkling of stars surrounding the blur.

A telescope of 8–12-in. aperture shows many of the brighter globulars well and allows some of the dimmer ones to be partially resolved into stars. In a large telescope (16–24-in.), at medium power (150×), a bright globular cluster is a truly awe-inspiring sight. Standing in line at a "star party" to see one of the bright globular clusters through a large telescope is worth the wait.

CHAPTER 9

Far Outside Our Solar System— Other Galaxies

As late as 1920, most astronomers still believed galaxies were spiral nebulas within our own Milky Way galaxy. However, within only a few years of that date, Edwin Hubble had made observations with the 100-in. telescope at Mount Wilson that proved that the Great Andromeda "nebula" is really quite remote from our galaxy, and thus a galaxy itself. Hubble's estimates of the distance to the Andromeda galaxy was 1 million light-years. Refinements since then have placed its true distance at about 2.3 million light-years.

Galaxies are sufficiently remote that they are difficult objects to observe in small telescopes. Despite this, they are among the most interesting objects to observe in the sky. Perhaps it's because you can point your telescope at the Andromeda galaxy and proudly tell your friends they're looking at an object as it was 2.3 million years ago. Maybe it's because of the incredible variety of galaxies there is to sample. Possibly it's the challenge of observing these remote, dim objects.

Whatever the reason, galaxies seem to have a permanent attraction for amateur astronomers. This chapter will get you started on what promises to be a lifelong but pleasant addiction.

9.1

How Galaxies Formed

Ours and all the other galaxies formed soon after the Universe began, between twelve and fifteen billion years ago. The Universe apparently began with an incredible explosion, and has been expanding ever since. At a time soon after this Big Bang, the matter in the Universe, which was almost exclusively Hydrogen and Helium, began to cool to the point that the gravity of the individual atoms was not overcome by the tremendous heat energy of the atoms. In short, the atoms were sufficiently cool to attract one another rather than bouncing off each other.

Local points of high density appeared all over the Universe, and these points attracted more and more material due to their gravitational pull. Eventually, enough material was attracted to form galaxies.

An infant galaxy must have been one vast nebula. At locally dense points within these nebulas, the first stars formed. Since there were only Hydrogen and Helium available, these stars contained no heavy elements, and the fusion reaction that powers stars could only produce relatively light elements.

Where do the heavy elements we find on Earth and in the Sun come from? When the first generation of massive stars had reached old age, they became supernovas. The pressure inside the supernova created many heavy elements, which were then blasted out into the nebulas surrounding the first-generation stars. Our Sun and other stars formed from these supernova-enriched nebulas.

Everywhere we look in space, there are galaxies. There may be 100 billion of them. The Milky Way galaxy is fairly typical of one type of galaxy, the spiral. For review, our galaxy is about 100,000 light-years across. Its disc is 2,000 light-years thick and its central bulge is 10,000 light-years thick. As far as its number of stars is concerned, our galaxy is typical: It has 100 billion.

Besides our galaxy, which is a normal spiral, there are three other types of galaxy.

9.2

Types of Galaxies

Spiral galaxies are arguably the most beautiful. They feature graceful spiral arms with contrasting dust lanes between the arms. They also feature a mixture of young and old stars, as there is still much gas and dust out of which new stars are continually forming.

Barred spirals seem to be a special variety of spiral in which each arm extends straight out from the center of the galaxy, and only begins to spiral around once it reaches the edge. The straight section on each side of the nucleus is like a bar, so the galaxy is called a barred spiral. Some SB galaxies, as they are called, are wrapped very tightly, while others are hardly wrapped at all.

Elliptical galaxies seem to be vast egg-shaped conglomerations of old stars with no recognizable structure, no spiral arms, and little or no gas or dust. They range in shape from those that are almost spherical to those that are almost cigar-shaped. Although they are of interest to astronomers, they are rather boring to observe. They just start out bright in the center and fade from there to the edge.

Irregular galaxies are just that—irregular in shape. They are made up of mostly bright young stars, much gas and dust, and some old stars. The satellite galaxies of our galaxy, the Magellanic Clouds, are irregular galaxies.

Soon after galaxies were discovered to be the incredibly distant objects they are, theories about how they attain their shapes have abounded. Most suggested that the various shapes represent different stages in the development of galaxies. Astronomers now believe that the different types of galaxies we observe were formed under different initial conditions.

A spiral will *always* be a spiral, and a barred-spiral will always be barred from being a regular spiral. Ellipticals are apparently condemned to being the old-folks' home for stars, with no new material from which to make new stars. And the irregulars? Well, they're just…irregular.

Even though most of them aren't observable by amateurs, no discussion of the types of galaxies would be complete without mentioning there are galaxies that don't fit into any of these categories.

Some of them are galaxies that look normal in visual light, but radiate enormous amounts of energy at radio frequencies. These are called radio-galaxies. Others look normal in visual light, but their output in infrared light is enormous. This type of galaxy was discovered by an astronomer named Seyfert, so they bear his name.

Finally, we come to quasars. These appear to be extremely remote objects, some as much as twelve to fifteen billion light-years away. If they are that far away, we must be seeing these objects as they were twelve to fifteen billion years ago, because it must

have taken the light that long to reach us. The distinguishing feature of quasars is that they are no bigger than our Solar System, yet they radiate as much energy as a hundred normal galaxies.

Recent observations have shown that quasars are located at the heart of galaxies, so we may be seeing an early form of galaxy that features a very active nucleus. What powers this nucleus is still unclear, although most models assume there is some sort of ultra-massive object at the center of a quasar. As the massive object sucks stars into it, we see their death throes as an incredible output of energy.

9.3
Groups of Galaxies

Our galaxy is part of a cluster called the Local Group. Among its members are our satellite galaxies (the Magellanic Clouds), the Great Andromeda galaxy and its satellites, the beautiful spiral galaxy M33 in the constellation Triangulum, and maybe a dozen more assorted minor galaxies.

Our Local Group is part of a supercluster of galaxies called the Virgo Supercluster, because its center lies in the constellation Virgo. Superclusters of galaxies range in size from as little as 100 million light-years across to as much as a billion.

Some astronomers now believe they have found evidence that superclusters of galaxies are arranged in vast strings throughout the Universe. It's unclear what significance this arrangement has. Although we'll never be able to see these "cosmic strings," completeness demands that they be mentioned.

9.4
Observing the Galaxies

Some galaxies have extremely high surface brightness, and can stand a lot of power. Others seem to be barely brighter than the background sky, and can only be viewed at extremely low power. This means you should always examine a galaxy at low power, then step it up until you no longer see anything new or any improvement over what you saw at a lower power. Most nebula filters I've tried seem to do little to enhance the view of a galaxy.

About the ALL-Sky Map for January

Possibly you received this book as a year-end holiday gift, and by the middle of January, the dust of the holidays, Boxing Day, and the bowl games has settled, and you've decided to actually keep one of your New Year's resolutions: to use this book to learn the sky!

My personal belief is you're much likelier to keep a resolution you want to keep, so I'll do my best to make this an enjoyable experience. Happenstance provides the brightest, easiest-to-learn constellations to help you get started!

Following the Path to January's Stars

If you don't already know, now would be a good time to learn which direction is north (see section 2.5.1). Exactly opposite north on your horizon is south, where we'll begin our trip through January's sky. Facing south, you should find a brilliant, blue-white star about a third of the way up from the horizon to the zenith. (If you live quite far to the north, this star will be closer to your horizon.) This is Sirius, the brightest star in our sky.

Sirius is also the brightest star in the constellation **Canis Major**—the Larger Dog, so it's often called the Dog Star. Take a moment to trace the outline of Canis Major. You can start by noting that the star that marks the dog's forepaws is about 5° to the right of Sirius. About 10° below and to the left of Sirius is a bright triangle of stars that forms his hind legs. Not a bad dog!

Just to make sure you've really found Canis Major, go over what you've just read and compare what you see with the outline I've drawn on the ALL-Sky for January. (Canis Major is the subject of the ZOOM map for February. If you want to know more about the Big Dog, jump ahead to February and read the description of its ZOOM map.)

Once you've recognized Canis Major, you can move on to one of the favorite constellations, **Orion.** This celestial giant is located above and to

the right of Canis Major. Despite the brilliance of Sirius, Orion is the dominant constellation in the winter sky. His belt, made of three bright stars in a row, and the brilliance of Betelgeuse at his right shoulder, and Rigel at his left knee, make it easy to imagine Orion as the mighty hunter. Indeed, Orion is often the easiest constellation for beginners to recognize, even though generations of scouts learned the Big Dipper first. The celestial equator passes just north of Orion's belt, so astronomers in Quito, Ecuador, or Nairobi, Kenya, would see Orion at their zenith. (Orion is the subject of January's ZOOM map, so I'll not dwell on it further.)

About 15° to the left, or east in the sky from Orion, you should notice a blue-white star that's not quite as bright as Sirius. This is Procyon, the brightest star in the constellation **Canis Minor**—the Lesser Dog. That this second of Orion's dogs is there at all appears to be caused by the ancients' need to create a constellation for such a bright star as Procyon.

Right above the (very) Lesser Dog, just east of the zenith, lies the constellation **Gemini**—the Twins. At their head lie the two bright stars Castor and Pollux. The ancients envisioned the rows of stars I've drawn as two young twin boys standing in a brotherly embrace. East of Gemini is **Cancer**—the Crab, a constellation that is more than a bit difficult to envision. When you're finding Gemini and Cancer, be sure to remember that these are two of the twelve constellations of the Zodiac, the constellations through which the Sun, Moon, and planets pass. To avoid being fooled by an extra star, be sure to check the mini-almanac for the month and year you're viewing. If a planet is predicted to be in one of these constellations, or in Leo, Taurus, Aries, or Pisces, you can find it by determining which bright star doesn't fit the pattern you find on the map. (The March ZOOM map depicts Gemini, Cancer, and Canis Minor.)

Just west of Gemini and due north of Orion lies a beautiful pentagon of stars. The southernmost in

ALL-Sky Map for January

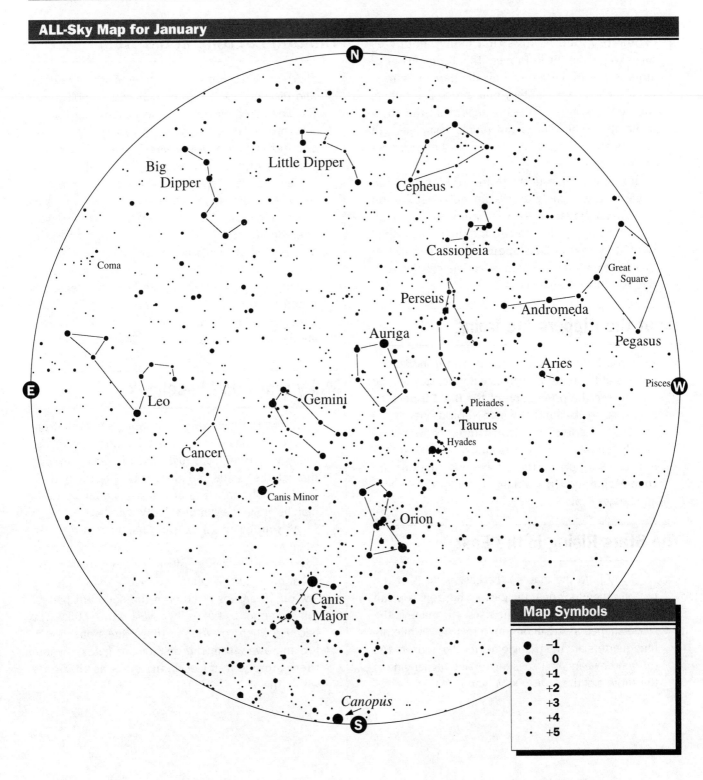

Map Symbols

●	−1
●	0
●	+1
•	+2
•	+3
·	+4
·	+5

this pattern is actually part of the constellation **Taurus**—the Bull. However, I've drawn the pentagon in its entirety, because that's the way almost everyone recognizes the constellation **Auriga**—the Charioteer. It's a nice pentagon, but I have a bit of trouble envisioning Auriga as a charioteer.

Now work your way back southwest of Auriga to a small, bright V-shaped group of stars. You'll know

you've found it if you see that the southern end of the V is a bright orangish star. You've found a group of stars called the **Hyades,** which make up the head of Taurus, the Bull. The rest of the bull is off to the east between Orion and Auriga. Since no one except an obsessive-compulsive classics scholar actually sees the bull, I've only outlined the Hyades. I'm betting that's what you'll see in the sky, too.

North and west of the Hyades, you'll find an exquisite group of stars that many neophytes mistake for the Little Dipper. This little cluster of stars is the **Pleiades,** technically part of Taurus. However, even though they only get co-star billing, the Seven Sisters of the Pleiades steal the show night after night. (Auriga and Taurus, complete with Hyades and Pleiades, are shown on the December ZOOM map.)

If you continue north from the Pleiades, you come to the vaguely arrow-shaped constellation known as **Perseus**—the Champion or Hero. Perseus's arrow points toward the slightly squashed W of stars that is **Cassiopeia**—the Queen in Her Chair. (Perseus and Cassiopeia are the headliners of the November ZOOM map.)

What the Dippers Are Doing

If you hold the ALL-Sky map so the northeast point is at the bottom, then face northeast, you'll be able to see the orientation of the **Big Dipper.** It's standing on its handle, which should ensure it's empty! To its left, the **Little Dipper,** however, is mostly full, as its bottom is almost parallel to the horizon. For purists, these two constellations are **Ursa Major** and **Ursa Minor**—Latin for the Greater and Lesser Bear.

The Stars Rising in the East

If you hold the map so the east point is at the bottom, then look near the eastern horizon, you can pick out **Leo**—the Lion as he rises. His mane is the sickle-shaped asterism on top in this view, while his hindquarters are the triangle nearer the horizon. (As the winter wears on, Leo is higher and higher, and is the subject of the April ZOOM map.)

The Stars Setting in the West

The nondescript constellations **Aries**—the Ram, and **Pisces**—the Fishes are sinking into oblivion in the west. To their right on the western horizon, an enormous "dipper" of stars made up of the constellations of **Andromeda** and **Pegasus** is standing on the horizon with its handle up. The ancients viewed Pegasus as a winged horse, but its distinctive almost-square shape is much easier to see, so I've drawn it as the Great Square of Pegasus. The edge distortion of the map has flattened the square a bit, as you'll notice when you actually see Pegasus in the sky. Since the Sun sets early these January nights, you can see all these constellations much better by going out at 8:20 PM with the December ALL-Sky map. (Andromeda and Pegasus are the subjects of the October ZOOM map.)

Milky Way and Miscellany

Sorry, but "miscellany" is not a constellation. This section deals with the fact that on this map, the Milky Way runs from just east of south, directly through the zenith, and plunges back to the horizon just west of north. The brightest stars on January's map lie along the winter Milky Way.

As you may have read in Chapter 8 of the text, when we look at the winter Milky Way, we're looking away from the center of our galaxy into a spiral section called the Orion arm. Note that there are few bright groups of stars in the southern part of the sky either to the east or west of the Milky Way because when you look in those directions, you're looking away from the disc of our galaxy, where there simply aren't as many stars as there are within the disc.

About the ZOOM Map for January

For January, we zoom in on **Orion,** the brightest and best known constellation in the sky. Since Orion lies on the celestial equator, there's almost no inhabited place on Earth from which it cannot be seen at some time during the year. The mythical character it represents is the heroic son of the Greek god Neptune.

Although he lived up to all the hype about mythical heroes, Orion was apparently also one arrogant son-of-a-god. He also dated the head cheerleader, Diana, but chased other goddesses. As word of his arrogance (and dalliance?) spread far and wide, Orion came to the attention of Juno, who sent a scorpion to sting him after hero-practice one day. After Orion died he was placed in the sky as a tribute to his heroism. So he wouldn't have to spend all eternity near his tormentor, the scorpion, he was placed exactly opposite that creature in the sky. They never appear in the night sky at the same time. (**Scorpius** appears on July's ZOOM map.)

The bright orange-red star that marks Orion's right shoulder is named Betelgeuse, which translated from Arabic means "arm-pit of the Great One." Betelgeuse is a "great one" itself, being a red supergiant star that varies in brightness from 0.4 to 1.2 magnitude. It takes a bit less than six years to go through one cycle of variability. When it's at its brightest, Betelgeuse is about 800 million miles in diameter. At its dimmest, it's "only" 500 million miles across. Betelgeuse is one of the largest of all known stars.

At its brightest, Betelgeuse puts out possibly 15,000 times as much energy at all wavelengths as our Sun. It's about 500 light-years from us, yet its brightness allows it to be one of the brightest stars in our sky. If it were as close to us as many other "bright" stars, Betelgeuse would dominate our night sky, and would also be easily visible in the daytime.

This single star would easily swallow up our inner Solar System well past Mars, yet Betelgeuse is possibly only twenty times as massive as our Sun. That means its mass is spread over an incredible volume of space. In fact, Betelgeuse is an incredibly hot, extremely tenuous sphere of plasma. It's almost an extremely brilliant—*nothing.*

Marking Orion's left knee is brilliant Rigel, a young, white supergiant star that's about 900 light-years distant. Whereas Betelgeuse is one of the largest of stars, Rigel is one of the most luminous. It puts out more than 50,000 times as much radiant energy as our Sun and shines at magnitude 0.2. Rigel, if it were as near as some other "bright" stars, would also be easily visible in daylight. It's about 40 million miles in diameter and masses fifty times the Sun. Sadly, such a star consumes its thermonuclear fuel very rapidly. Our Sun will outlive Rigel by many lifetimes.

Rigel has a companion star orbiting it. It's bluish, of magnitude 6.7, and is 9" distant from Rigel itself. Even though it's quite far from Rigel, the extreme difference in magnitudes makes it a challenging double to split. On a steady night, use at least a 3-inch refractor at high power.

The **Great Orion Nebula** is the *pièce de résistance* of emission nebulas. I first saw this object through a 60-mm refractor from within the city. That first view was magnificent. I've seen the Orion Nebula through my 20-in. (508-mm) reflector telescope, and it was truly awe-inspiring. However, that view doesn't diminish the view I had as a child through that 60-mm refractor. This object is wonderful in all kinds of instruments.

You can find the Orion Nebula by finding Orion's sword hanging just below his belt. The sword is made up of three stars. If you look at the middle one under fairly dark-sky conditions you'll notice it's a bit misty or fuzzy. Don't "adjust your set" or blame your eyes, what you're seeing is really the Orion Nebula. It's *that* bright.

When you view the Orion Nebula with even a modest telescope, you'll immediately notice that the middle star is enveloped in a vast, greenish-tinged bubble of nebulosity. Look a little closer and you'll notice that what you thought was the brightest star in the nebula is actually four!

ZOOM Map for January

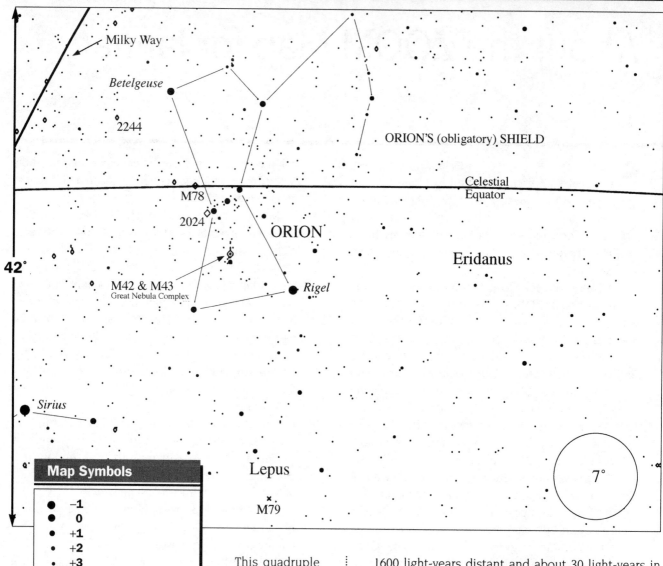

Milky Way

Betelgeuse

2244

ORION'S (obligatory) SHIELD

Celestial
Equator

M78

2024

ORION

Eridanus

42°

M42 & M43
Great Nebula Complex

Rigel

Sirius

Lepus

7°

M79

Map Symbols

●	−1
●	0
●	+1
•	+2
•	+3
·	+4
·	+5 and +6
∞	**Galaxy**
×	**Globular Cluster**
◦	**Galactic (open) Cluster**
◈	**Nebula**

This quadruple star is known as the **Trapezium.** All of its stars, and several more much dimmer companions, were formed out of the same cloud of gas and dust at the heart of the Orion Nebula. Mathematicians have not yet developed the techniques to describe accurately the orbits of so complicated a system. However, we don't need to know how they move to enjoy viewing this wonderful star system.

The Orion Nebula is merely the center of a vast cloud of gas and dust that makes up the center of the figure of Orion. This cloud may have spawned great Rigel and several other hot, white supergiant stars nearby. The Orion Nebula itself is possibly 1600 light-years distant and about 30 light-years in diameter. There appears to be enough gas and dust to form ten thousand stars like our Sun. Recent observations by the Hubble Space Telescope seem to have observed many tiny globules in the cloud, each a potential new star. Other observations indicate the presence of protostars that have already begun their collapse into true stars. The Orion Nebula is truly a cosmic nursery.

Binoculars allow you a view of the entire Orion Nebula and its mate, called M43, just to the north. A small telescope, used at low power, reveals the subtle whirls and streamers within the outlying portions of the nebula. Several observers have suggested pointing your telescope to the east of the nebula and letting it slowly drift into your field of view, all the while examining the outer portions of the nebula. I've tried that, and it does allow me to see more of the subtle outer reaches.

(If you have a clock drive, you'll want to turn it off for this exercise.)

Increasing the power allows you to examine the brighter area near the Trapezium itself, especially the "notch" just north of the famous quartet. This is a dark nebula between us and the emission nebula. Many observers, as they become experienced, begin to see subtle pink and purple tints near the center of the nebula. Little kids, whose color vision is much more acute than adults', have often viewed the Orion Nebula through my 20-inch and immediately noticed the pinks and purples, much to my chagrin. Regardless of your age or observing prowess, the Orion Nebula will never fail to please!

Besides having the biggest star, the brightest one, and the most wonderful nebula in the sky, Orion does feature other objects, as do the surrounding constellations, so here's a round-up of them:

I've added Orion's shield to the constellation figure that appears in the ZOOM map, mostly to provide you a path to follow with your binoculars or small telescopes. Several interesting galactic clusters lie north of Betelgeuse, and at least one along his shield. Can you find them? If so, mark their position on the map. After all, these are NOT sacred pages.

Just east of Orion's belt lies another nice nebula, **NGC2024.** Leading south of it is a faint sheet of nebulosity that's partially blocked by a dark nebula in the shape of a horse's head. This is the famous **Horsehead Nebula** that's featured in so many photographs of the region. It's relatively easy to photograph, and extremely difficult to see visually. I've seen it only half a dozen times in 35 years of observing. Another nebula, **M78**, lies just 2° north of NGC2024.

At the extreme southern edge of January's ZOOM map lies a "hairy" star, the globular cluster **M79**. It's appropriate that this star is "hairy," because it's found in the constellation **Lepus**—the Hare. M79 lies a bit more than 100 light-years from us and shines at magnitude 8.3. Since that light is shared by many stars, M79 is not one of the finest of globulars. It's worth a look, though.

About 10° east of Betelgeuse lies the galactic cluster **NGC2244** in the constellation **Monoceros.** This cluster is often visible to the naked eye and is also a nice object in binoculars. Surrounding the cluster is a wreath of faint nebulosity that's known as the **Rosette Nebula.** As with many nebulas, it's much easier to photograph the Rosette than to see it. However, I've often glimpsed it in binoculars, and a nebula filter makes it a relatively easy object. If you decide to try viewing the Rosette, the worst that will happen is that you'll only see a nice open cluster. HINT: The Rosette is BIG, about 80', or 1.33° in diameter. Use a very low-power eyepiece in your telescope to observe it.

Although they appear at the southeastern edge of January's ZOOM map, **M50** and **M41** are also shown on February's. I'll tell you about them in that month's commentary.

About the ALL-Sky Map for February

February finds the winter Milky Way and its bright constellations beginning to slide over to the western horizon as they prepare to disappear into the Sun's glare in the early spring. Also, there's a relative void of bright, easy-to-see constellations in the southern sky east of Canis Major. February's sky is a celestial bore—NOT! There's still lots to see.

The Path to February's Sky

The easiest way to find the constellations of the western half of February's sky is little different than the path for January. You must first know which direction is south. Looking a bit west of

Map Symbols

●	−1
●	0
●	+1
•	+2
·	+3
·	+4
·	+5

south, and about 30° above the horizon, you'll find Sirius, the Dog star, the brightest star in our sky. You'll have little difficulty tracing out **Canis Major,** with one bright star west of Sirius as his forepaws and the bright triangle southeast as his hindquarters.

It takes little imagination to imagine Canis Major standing up on his hind legs looking up at his master, **Orion**—the Hunter. With the three distinctive stars making up his belt, the orange-red

Betelgeuse marking his right shoulder, and bluish-white Rigel his left knee, Orion is the easiest constellation to find.

To the east of Orion, you'll find the bright bluish-white star Procyon, which (somehow) marks his other faithful hunting companion, **Canis Minor.** Just above Canis Minor, near the zenith lie Castor and Pollux, the (almost) twin stars that mark the heads of **Gemini**—the Twins. When NASA decided to build a two-seat spacecraft intermediate in size between the original Mercury capsule and the Apollo moon craft, it was named Gemini.

Just above Orion, northwest of Gemini, you'll find **Auriga**—the Charioteer. The southernmost star in the pentagon I've drawn is really within the boundary of the constellation Taurus. However, most stargazers I know see the pentagon as Auriga, so I've drawn it that way, too.

To the northwest of Orion, over his left shoulder, you'll find the V-shaped **Hyades,** which make up the head of **Taurus**—the Bull. His horns are the two bright stars to the east. And, again, the top star has been stolen to make Auriga's pentagon. On the other side of Taurus from Orion you'll find the lovely **Pleiades**—the Seven Sisters.

Since the ecliptic runs between the Pleiades and the Hyades, there are often extra stars found there: the planets. Also, the Moon can pass as much as 5° north or south of the ecliptic, so it can occult the Pleiades or the Hyades. Watching such an occultation through binoculars can be a beautiful sight.

If you continue along the winter Milky Way past the Pleiades, you come to **Perseus**—the hero, and **Cassiopeia**—the Queen in Her Chair. **Cepheus**—the King of Ethiopia is next along the Milky Way, lying next to his queen just under the North Celestial Pole.

Both **Andromeda** and **Aries,** although above the horizon, are too far down in the murk to try to learn this month. You'll have to wait until next fall to see them again in the evening.

An Eastern Branch in the Path

So far, you've only learned the western half of February's sky. That part of the sky is setting, so it should get first attention. Now you can learn the constellations of the eastern half of the sky at your leisure, because those constellations won't set for many hours.

Start by going back to the zenith and finding Gemini. Remember, it's the constellation of the Twins just north of the bright star Procyon. Moving East, you come to **Cancer**—the Crab, another of the

constellations of the Zodiac. It's seldom that anyone really looks at Cancer and says to herself, "Wow, that looks just like a crab." However, the ancients didn't get out much, so they spent a lot of time inventing constellations like Cancer, and now we're stuck with them. If you're out on a dark night, you might be able to see a "fuzzy glow" at the heart of Cancer. See the March ZOOM map for a thorough description of this cluster of stars known as the "Beehive."

Between Cancer and the eastern horizon lies one of the easier constellations to identify, **Leo**—the Lion. Leo dominates what appears to be an otherwise empty portion of the sky, so he should be easy for you to find. The lion's head and mane are formed by a sickle-shaped group of stars that many call the "backwards question mark." The triangle of stars you see below the sickle are his haunches. I've always envisioned Leo resting on his haunches in a Sphinx-like pose, his forepaws ahead of him.

The February sky marks the reappearance of one of the summer's bright stars, Arcturus, the brightest star in the constellation **Boötes**—the Herdsman. (Boötes is pronounced Bo-OH-tees, by sounding both O's and slightly accenting the second.)

Just below Leo, you'll find a dim Y-shaped group of stars "standing" on the eastern horizon. It's **Virgo**—the Virgin. Why this "Y" was seen as a virgin I can't explain.

To the left of Virgo, near the southeastern horizon, you'll find a kite-shaped group of stars known as **Corvus**—the Crow. Corvus makes a much better kite than a crow.

What the Dippers Are Doing

Facing north, you should be able to see the **Big Dipper** above and to the right of the North Celestial Pole, and Polaris, the star that's very near it. Once you've found Polaris, follow the arc of stars down and to the right until you find the bowl of the **Little Dipper.** Don't be discouraged if you can't find the Little Dipper the first time you try. It's not nearly so easy to find as the big guy.

About the ZOOM Map for February

For February, we zoom in on Canis Major—the Greater Dog, and the southern winter Milky Way that meanders just east of his tail. This is a wonderful area to tour with binoculars on a cold winter evening. However, before we take a look at the bright deep-sky objects found on this ZOOM map, it's appropriate to give the brightest object in that part of the sky its due—Sirius, the brightest star in the sky.

It's really hard to miss Sirius; its blue-white glare reminds even the casual observer of a celestial diamond. It's at the head of **Canis Major,** whose constellation figure makes a pretty good dog, which makes Sirius even harder to mistake for any other bright star.

Sirius shines at magnitude −1.4, and can easily be seen with a telescope or binoculars on a really clear day. (When I was ten or eleven, I remember seeing Sirius on many clear late-winter afternoons well before sunset. After trying to convince a few adults of this, I gave up. Now it's 36 years later, and it's the kids who can see everything, while I'm busy trying to find my reading glasses, or fixing the pair I sat on....)

At 8.7 light-years, Sirius is the fifth-closest star known. Part of its apparent brightness is how close it is. However, it's also intrinsically bright, 23 times brighter than the Sun. Its surface temperature is about twice the Sun's—about 10,000° K, and the temperature at its core is 20 million°K! It's nearly twice as big as the Sun, and masses more than twice the Sun's mass.

Although we know of no planets circling Sirius, we do know of a companion star, Sirius B, that takes fifty years to complete one orbit. Before this companion was discovered, its existence was predicted mathematically. Since the time of Christ, Sirius has moved on the Celestial Sphere about 44", or about one and a half times the diameter of the Moon. Many measurements of the position of Sirius were made over a period of several hundred years after humans began making serious observations of the heavens. In the middle of the last century, the

great German mathematician Friedrich Bessel noticed that Sirius's path on the Celestial Sphere was not a straight line, but had "bumps" at fifty-year intervals. Although astronomers tried to see Sirius B for some years after Bessel's prediction, it remained invisible until found by accident in 1862.

The great American telescope maker Alvan G. Clark was testing a new 18.5-inch refractor he had just made for Northwestern University. He pointed it at Sirius, and easily saw Sirius B! (Needless to say, the telescope passed the test.)

The reason Sirius B had remained so elusive was that it is only an 8.7-magnitude star. So, it's 10.1 magnitudes, or about 10,800 times dimmer than Sirius A. Since it can be as close as 2' from Sirius A, Sirius-"pipsqueak"-B is often lost in the glare of its flashier sibling. I've seen it exactly once, through an excellent 10-inch Clark refractor, but that was when its separation from Sirius A was quite a bit farther than it is now. Also, the seeing was almost as good as I ever knew it. In the decade of the nineties, splitting Sirius would be an incredible feat, as its separation will be at or near the minimum during the whole decade.

Once they knew the orbit of the pair, astronomers could determine that the total mass of the system is 3.33 times that of our Sun. From its spectrum, they were able to estimate the mass of Sirius A as 2.35. Thus, dim Sirius B masses 0.98, or almost the same mass as our Sun. Yet its absolute magnitude is much less than our Sun, implying that Sirius B is an extremely cool star.

Once a spectrum was obtained, however, the surface temperature of Sirius B was found to be a hot nearly 9000° K, considerably hotter than the Sun. What gives?

The answer is that Sirius B is extremely small. It's a star that's exhausted all its nuclear fuel and is no longer generating enough energy to keep it "inflated." Its own gravity has collapsed it to planet size—only 19,000 miles in diameter—and is also responsible for heating it up. It's called a white

ZOOM Map for February

Celestial
Equator   Monoceros

M48

M50

M46
M47   *Sirius*

Puppis   M41

M93   Canis
Major

Antila

49°

Milky Way

Map Symbols

●	−1
●	0
●	+1
●	+2
·	+3
·	+4
·	+5 and +6
∞	Galaxy
×	Globular Cluster
◇	Galactic (open) Cluster
◈	Nebula

7°

dwarf because its color is very white and it's about as small as stars get.

Eventually, Sirius B will radiate all its remaining energy into space and will become a black dwarf, the cinder of a dead star. It will then be a menace to navigation, as the original *Enterprise* demonstrated in one episode of "Star Trek." Unfortunately, in the real world, passing very close to a black dwarf doesn't propel you and your crewmates back in time. But it will give you a king-size headache!

Sirius dominates this portion of the winter sky, but there are many other interesting objects for your viewing pleasure. With one surprise exception, they're all galactic, or open clusters. The first is **M41,** a lovely object just about 4° south of Sirius. M41 should be easy to find, as it's usually visible with the naked eye on a dark night, despite Sirius's glare.

M41 is a big, scattered cluster about as wide as the full Moon. Its centerpiece is a seventh-magnitude red star. That star's 100 companions tail off until the dimmest is about magnitude 13. M41 is about 2400 light-years away, and it's 20 light-years across.

M50 lies about 8° north, and 3° east of Sirius in the constellation **Monoceros**—the (totally invisible) Unicorn. This cluster is somewhat "tighter" than M41, and seems to have fewer stars, but one of them is reddish, just like its neighbor to the south. M50 is about 3000 light-years away, and is about 10 light-years across. Since its total magnitude is 6.3, you might be able to see it with your unaided eye on a dark night, especially if you block the light of Sirius behind a tree or building.

You'll find **M46** a bit over 14° east of Sirius. (That's twice the 7° field of the typical 7×50 binocular.) This cluster is big, and its 150 stars are between magnitude 10 and 13, so it usually appears as a cloud about the diameter of the full Moon in binoculars or a small telescope. If you view M46 through a larger telescope, it presents a field that looks like tiny diamonds have been scattered about the sky. It's one of my favorites.

M46 is about 5400 light-years away, and about 30 light-years across. There's a small planetary nebula, **NGC2438,** just at the northern edge of M46 that appears to be associated with the cluster. It's actually some 2000 light-years closer, and just happens to be almost exactly in our line-of-sight to the cluster. It will take a 4-inch telescope and a dark, steady night to see this little planetary. A larger scope makes it a relatively easy object.

A degree and a half west of M46 is **M47,** a closer, brighter, and sparser cluster. The brightest stars in M47 are of fifth and sixth magnitude, so they're between 40 and 100 times brighter than those in M46. M47 is only a third the distance of M46, which is one reason its stars appear brighter. The other is that its brightest stars are intrinsically much brighter than those in its more distant neighbor. M47 also features several nice double stars.

To the east of the tail of Canis Major, and about 9° south of M46/M47 lies **M93,** a compact cluster of about 60 stars, all within a diameter of 20", or about two-thirds the diameter of the full Moon. The brightest stars are of magnitude 7, and they trail off to magnitude 12. Since their combined magnitude is about 6.2, you should be able to just detect M93 without optical aid on a clear, dark night. Again, I advise you to block the light of Sirius before trying this. M93 is nearly twice the distance from us as M47, about 3400 light-years, and its diameter is 18 light-years.

Finally, you can find **M48** by sweeping 16° northeast of M46/M47. M48 is a group of stars that total magnitude 5.9, so it's another candidate for naked-eye visibility. M48 is big, nearly twice the diameter of the full Moon, so you'll want to use your lowest power to see its 50 or so stars.

About the ALL-Sky Map for March

As we dodge late winter snowstorms or early spring thunderstorms, a different kind of sky presents itself in March. The bright constellations of the winter Milky Way now repose along the western horizon, while their summer replacements are beginning to peek over the eastern horizon. One result is a dimmer sky as we look almost at right angles to the plane of our Milky Way galaxy. Another is a pair of paths to find the stars of March.

Following the Western Path to the Stars of March

If you want to learn the stars of the winter Milky Way, do so either early in the month, or a bit earlier than the standard time of 10:20 P.M. This will give you some time to trace them all before they slip over the horizon.

As usual, we start with Sirius, which you'll find as the brightest object low in the southwest. In March, the ecliptic meets the horizon well north of due west, so you should have no trouble confusing a planet for Sirius. The Larger Dog—**Canis Major,** of which Sirius is the brightest star, appears to be loping along the southwestern horizon, faithfully following his master, **Orion**—the Hunter.

Orion almost appears to be leaning over toward the horizon, just south of due west. The key to seeing him is to remember that the three stars of his belt point back toward Sirius. Also, a line running from Rigel, the bright star at his left knee, through

ALL-Sky Map for March

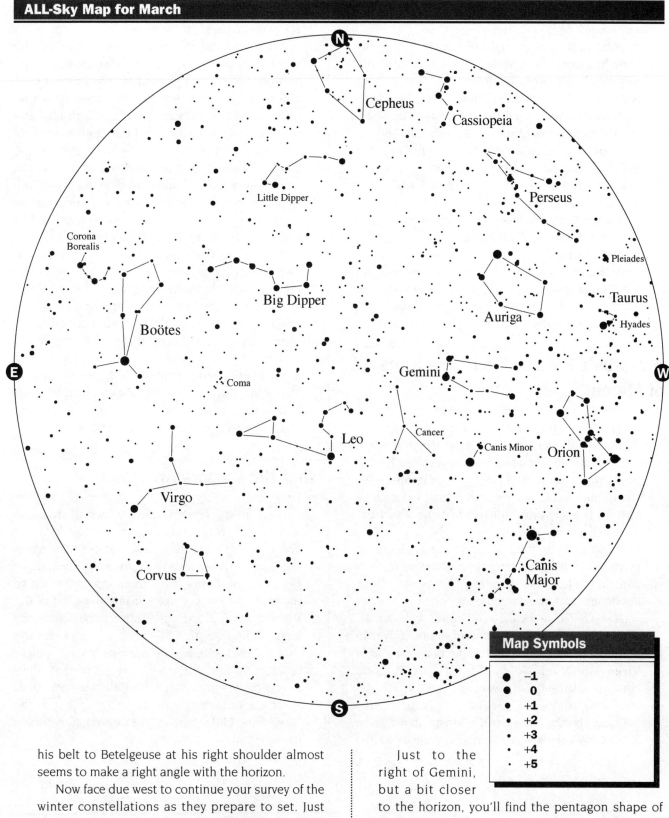

Map Symbols

●	−1
●	0
●	+1
•	+2
·	+3
·	+4
·	+5

his belt to Betelgeuse at his right shoulder almost seems to make a right angle with the horizon.

Now face due west to continue your survey of the winter constellations as they prepare to set. Just above leaning-Orion's head you'll find the figure of **Gemini**—the Twins. Two lines of stars seem to run straight toward the zenith. Before they reach it, however, they culminate in Castor and Pollux, stars that mark the heads of the Twins.

Just to the right of Gemini, but a bit closer to the horizon, you'll find the pentagon shape of **Auriga**—the Charioteer, with bright Capella at the upper right-hand position of the pentagon. Immediately below Auriga, the V-shaped **Hyades** point directly at the horizon just a bit north of due west. Since the Hyades are the head of **Taurus**—

the Bull, in March he seems intent on goring the western horizon!

A bit farther to the right of the Hyades lie the **Pleiades.** Once you find this exquisite, bright little galactic cluster, you'll never tire of seeing it. Once you've found the Pleiades, move a bit toward the northwest, where you'll find **Perseus**—the Hero. The top ragged row of stars in the figure of Perseus is almost parallel to the northwestern horizon, and the other row tilts down toward it. When he's near this horizon, I see Perseus leaning over toward his mother, Cassiopeia.

To see **Cassiopeia**—the Queen in Her Chair, look near the northern horizon only a bit to the west of due north. Her outline now makes a slightly squashed W. Completing this branch of the path, we find **Cepheus**—the King of Ethiopia. He's a rather dim but recognizable pentagonal figure wedged into the sky between the pole and the northern horizon.

Arcturus's Path to the Rising Stars of March

The fourth-brightest star in the sky, Arcturus, dominates the eastern half of the evening sky in March. This −0.1-magnitude yellowish-red beauty marks the lower part of the figure of **Boötes**—the Herdsman. You'll find Arcturus almost exactly due east in the sky and one-third of the way from horizon to zenith, or at 30°.

As we see him these March evenings, Boötes is pretty much lying on his side, which does little to conceal the fact that our intrepid herdsman has an enormous trunk and tiny, little legs.

If you're facing the eastern horizon and look 30° to the left and a bit lower toward the horizon than Arcturus, you'll come to Spica, the brightest star in **Virgo**—the Virgin. Virgo is the Y-shaped group between Spica and Leo's haunches.

If you continue to circle along the horizon to the left, you find the kite-shaped asterism known more formally as **Corvus**—the Crow, which is easy to find

about 15° to the left of Spica and just a bit higher above the horizon.

When you get to Corvus, you've reached the south "end" of the path. You could go on around the horizon, but to do so would take you into a vast, dim group of constellations only their inventors could recognize. Instead, face due south now and lean back as you look toward the zenith. Just a bit south and east of the zenith, you should find the figure of **Leo**—the Lion. His head and mane are formed by a sickle-shaped group of stars and he's facing to your right as you look south. He's resting on his forepaws and haunches, which are defined by a triangle of stars about 15° east of the sickle.

Continuing our animal motif, Leo is gazing at **Cancer**—the Crab, another constellation of the Zodiac. Cancer is barely recognizable as a crab, but because I decided to include drawings of all the twelve constellations of the Zodiac, I had to do something with this "crab." If you're viewing from a dark sky, you'll possibly notice a fuzzy glow just north of the center star of the figure of Cancer. That's the Beehive Cluster, about which you can learn more by reading the commentary on the March ZOOM map on the following pages.

What the Dippers Are Doing

If you're still facing south and looking up at Leo and Cancer, rotate 180 degrees so you're facing north. Leaning waaaaay back, you can once again see Leo, only he's upside down now. Look about two-thirds of the way from the northern horizon to the zenith, about 60° up. Cruising there is the **Big Dipper,** which you see in position to be continually empty. The two stars in the "bowl" opposite the "handle" of the Dipper are pointing almost straight down at Polaris, the magnitude 3 star that's less than a degree from the North Celestial Pole. From Polaris, follow the arc of stars to the right to the bowl of the **Little Dipper,** just beginning to dribble its contents onto its handle.

About the ZOOM Map for March

The reality of eating fruits and vegetables is that, when strawberries are in season, you eat them until they come out your ears, followed by corn, beans, and tomatoes, and so on. It's the same with observing the sky. When our Milky Way galaxy is "in season" you observe those objects that are its natives. Since there are possibly a thousand galactic clusters tucked away within our galaxy, you can expect to observe a lot of them when the Milky Way is high in the sky.

So it is with this last month of concentrating on the winter Milky Way. We'll observe a group of galactic clusters, just as we did for February. However, the first objects we'll observe will be some interesting stars.

You've already learned how to find **Gemini,** and its twin stars, Castor and Pollux. Castor is the northernmost twin, or on the right when you envision these bright stars as the heads of the Twins. To the naked eye, it's a pure white star of magnitude 1.6. However, once you examine Castor in a moderate-sized telescope, it's quickly evident Castor is a double star. If you look really close, under good conditions, Castor can be found to be triple!

Two of its components, Castor A and Castor B, shine at magnitude 2.0 and 2.9, respectively. This pair is separated by approximately 3", and their separation is widening for the entire decade of the nineties. They combine their light to appear to the unaided eye as a single star of magnitude 1.6. Almost any telescope worthy of the name can resolve A and B at medium power. Castor C is a red dwarf star of magnitude 9 that's about 73" south of A and B. Since it's so dim and so far from A and B, Castor C is not immediately evident as part of the system, and only a large telescope will reveal its red color to the human eye.

Castor is an extremely important star because it was the first double star observed whose components could be seen to orbit one another with each passing year. This simple fact was crucial to early astronomy, because it proved to

astronomers that the law of gravity not only applied to objects in our Solar System, but to objects in at least one other system as well. Once other true binary stars were observed to be revolving around one another, it became apparent that the law of gravity works the same way seemingly everywhere we look. Astronomers now accept as "gospel" that all the laws of physics apply the same way everywhere in the Universe, and this very powerful notion got its start with the early astronomers' observations of Castor A and B. (Sorry about the lecture, but this is "Big Stuff"!)

Castor's story doesn't end there. Once astronomers began making very accurate spectra of Castor A, B, and C, they noticed slight irregularities in those spectra that could only be caused by each of the three components being itself a binary—for a total of six stars in the system!

(The irregularities they noticed were that the positions of the spectral lines were shifted due to the orbital velocity of each star in the pair. The star that was approaching us "squashed" the waves of light, and the one that was receding "stretched" the waves. What was observed was the slight change in frequency, and thus the position of the spectral lines.)

Even more interesting is that each component of each pair is nearly identical. Each pair is also in a very "close" orbit. The longest orbital period is about nine days, for Castor A. Castor B's pair takes a bit less than three days to complete one orbit, while Castor C's pair does it in a dizzying twenty hours! To get a handle on the scale of this amazing system, consider this exercise:

Let's journey to scenic Dodger Stadium here in my adopted home sprawl of Los Angeles. We'll just shrink Castor A1 and A2, as we'll call them, to about an inch in diameter. We'll place them 2.5 inches apart, say, on either side of a baseball, and just set them on home plate.

We'll shrink Castor B1 and B2 the same way we did A1 and A2. They'll end up being 0.75 inches in diameter, and about 1.9 inches apart. To keep them

ZOOM Map for March

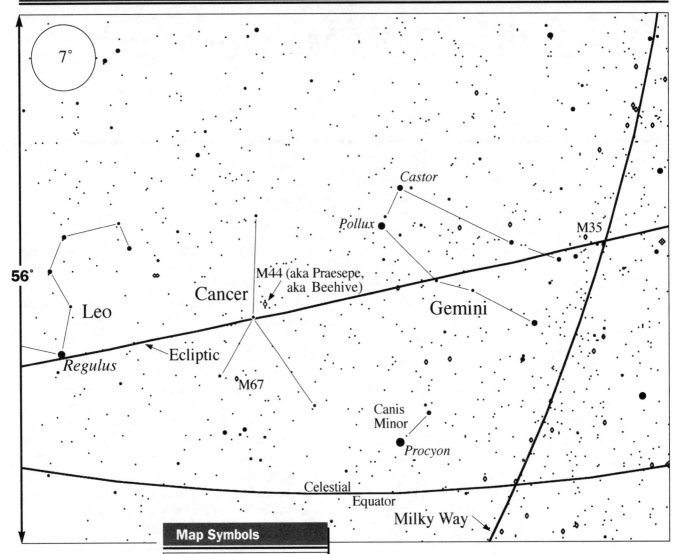

Map Symbols

●	**−1**
●	**0**
●	**+1**
•	**+2**
•	**+3**
·	**+4**
·	**+5 and +6**
∞	**Galaxy**
×	**Globular Cluster**
◦	**Galactic (open) Cluster**
❖	**Nebula**

to scale, we'll put B1 and B2 just 17 feet beyond the center field fence, approximately where the flag pole is, 412 feet from home plate.

So far, we've just gotten Castor A1 and A2, and Castor B1 and B2, the close members of the sextet, placed in our demonstration. What of the pipsqueaks, Castor C1 and C2? They're each about 0.35 inch in diameter, and 2.9 inches apart. Keeping C1 and C2 in scale requires that we place them 4900 feet from home plate! To do that, we must go past the farthest reaches of the parking lot, over the hills beyond

that, past Elysian Park, and into the median strip of the Golden State Freeway at Elmgrove Avenue!

On the same scale as the Castor System, our Sun would be 0.5 inch in diameter. If we placed the Sun at home plate with Castor A1 and A2, Earth would be about 13 feet in front of home plate, less than a fourth of the way to the pitcher's mound. Our home planet would be the size of the *thickness* of the paper on which this book is printed. Jupiter, the dominant planet in our system, would be 67 feet from home plate, barely past the pitcher's mound. Pluto would be 522 feet from home plate, out in the VIP parking lot.

Pollux, Castor's twin, is rather boring by comparison, although its name does mean something akin to "boxer" in Greek. It's a yellowish star of magnitude 1.2. Since it's yellowish, Pollux is a relatively cool star, at about 4500° Kelvin. Pollux is about 35 light-years away, so it's just beyond the standard distance for determining intrinsic brightness. This means its intrinsic brightness and its apparent brightness are almost the same. Although Pollux won't ever challenge the Rigels of this Universe, it is a respectable thirty-five times brighter than the Sun.

Of course, you'll want to observe Castor as a double star. Why not take a look at all of the bright stars that make up the figure of Gemini? There are some other doubles lurking there that space won't permit me to list. There are also numerous galactic clusters strewn around the Twins, making a tour of Gemini with binoculars a real treat.

The best of the clusters is **M35,** which lies 3° mostly west and a little north of the Twins' left foot. This is a nice scattering of 100–150 stars that combine their light to make M35 occasionally visible to the naked eye on a clear, dark night. It's about a half a degree across, so, at its distance of 2000 light-years, it's maybe 30 light-years across.

About 0.5° southwest of the center of M35 lies another galactic cluster, **NGC2158.** (To eliminate "cluster-clutter," I've *not* marked its position on the ZOOM map.) A small telescope may only show 2158 as a faint smudge at the edge of M35. However, a moderate-sized instrument will reveal it as a fine galactic cluster in its own right, although it's more concentrated than the typical "open" cluster. Since 2158 is 16,000 light-years away, or eight times the distance of M35, it provides an interesting view of two similar objects in the same field that are at very dissimilar distances.

Cancer has few stars of interest to a beginning stargazer. However, it does feature one of the finest galactic clusters in the sky, **M44.** Known as the "Praesepe," or "Beehive," this cluster is so big and bright that, even though the naked eye cannot resolve any of its individual stars, they combine to make it an easy naked-eye object. It was known to the ancients as the "little cloud."

Galileo was the first to point a telescope at the "little cloud." He resolved it into a myriad of stars, all "buzzing around each other." Hence the name. There are about 200 "bees" in the "hive," which is about 1° in diameter. At a distance of about 500 light-years, M44 is one of our closer neighbors among the clusters.

When you observe M44, keep the power low, or you'll end up seeing only a few of its stars at a time. This cluster really rewards the binocular viewer, who may actually get a better view of it than when seen through a telescope.

About 9° south of the Beehive is another bright "open" cluster, **M67.** This cluster is farther and more compact than the Beehive, so it can stand some power. It's about 2500 light-years away, and has possibly 500 stars, many of which are not detectable in a small telescope.

Despite its status as one of Orion's hunting dogs, **Canis Minor** has no objects worthy of observation in a small telescope. Even bright Procyon, which has a close, dim companion, is so much brighter than its companion that it is almost never observed. Its other companions are so far away from Procyon that they are of no interest to observers. For the record, Procyon, along with Sirius, is one of the five closest stars that can be seen with the naked eye. Its magnitude is 0.4.

About the ALL-Sky Map for April

Between the thunderstorms of spring, I hope you can find a night or two to study April's sky, an interesting interlude between the sky dominated by the winter Milky Way and that dominated by the summer Milky Way. Some folks think the evening sky in spring is boring. True, it's not as flashy as the winter or summer sky, but its subtle beauty is one I've come to appreciate. I think you will, too.

Since the Milky Way has "cleared off" to the horizon, we're looking out of our galaxy. Sure, all the

Map Symbols

●	−1
●	0
●	+1
•	+2
·	+3
·	+4
·	+5

stars, clusters, and nebulas we see in this view are part of our galaxy, but there's very little gas and dust to obscure our view MUCH farther out to other galaxies. And, as you'll see, there are a LOT of other galaxies!

The Western Path for April

Begin by standing, facing south, and craning your neck until you're looking almost at the zenith.

With little effort, you should recognize **Leo**—the Lion just southwest of the zenith. It helps to remember that Leo is one of the bigger figures. He's made up of a sickle-shaped head and mane to the west, and his haunches are formed by a nice triangle of stars to the east, so he's facing the western horizon.

West of Leo's mane about 20°, you may notice there's a "cloudy" or "misty" patch of sky, especially if you're out away from city lights. This is the galactic cluster **M44,** the "Beehive" that marks the position of **Cancer**—the Crab much better than its rather feeble outline figure.

A bit south and mostly west of the Beehive lies the brilliant blue-white star Procyon, the dominant star in **Canis Minor**—the Lesser Dog. There's really only one other bright star in Canis Minor.

You've probably slowly turned left to follow the path. If so, you're now facing the western horizon. **Gemini**—the Twins are standing on that horizon. Their brightest stars, Castor on the right, and Pollux on the left, lie 20° above the horizon and their bodies are defined by the parallel lines of stars that stretch to the horizon.

Speaking of stretching, you should compare the figure of Gemini found on the April ALL-Sky with what you see on the February and March maps. Notice that the Twins seem to get taller and taller as they move toward the western horizon. This stretching is caused by the distortion caused when we try to project a hemisphere, the sky, onto a flat surface, the pages of the book.

If you continue on around the western horizon toward the north point, you'll find **Auriga**—the Charioteer, the pentagon-shaped figure captained by the bright star Capella. Even farther around, and closer to the horizon is **Perseus**—the Hero, who's probably lost in the gloom. **Cassiopeia**—the Queen in Her Chair and **Cepheus**—the King of Ethiopia lurk beneath the North Celestial Pole.

I took you on this tour partly for the sake of completeness and partly to point out something important. From most of the places where you're likely to buy this book, the temperate regions of the northern hemisphere, Cassiopeia and Cepheus *never* go below the horizon. They disappear at dawn and reappear at dusk, but they're always in the sky, even in April when they just graze the northern horizon below the pole. In techno-speak, these are circumpolar constellations.

What the Dippers Are Doing

You started near the zenith and circled around to the left toward the western horizon and then continued on to face due north to notice that Cassiopeia and Cepheus, being royalty, refuse to go below the horizon. Now look up from the north point about two-thirds of the way to the zenith to find the **Big Dipper.**

Between Cepheus and the Big Dipper, you'll find the **Little Dipper,** with Polaris marking both the end of its handle and the North Celestial Pole. Polaris should be easy to find, since the two stars in the end of the bowl of the Big Dipper point almost directly to it. Remember, the angular distance is 28°, or about three fists away.

The Eastern Path to April's Sky

Since you've just found the Big Dipper, you can follow the arc of its handle around to Arcturus, a fiery reddish-yellow star whose brightness is exceeded by only three others in the sky. Arcturus anchors **Boötes**—the Herdsman, whose figure resembles the super-wide ties worn by used car salesmen in the seventies. Just to make sure you're looking where you want to, check to see that you're facing east and that the star you believe is Arcturus is about two-thirds of the way up from the horizon to the zenith. Well to the left of Arcturus and closer to the horizon lies a bright white star, Spica. This beauty is the brightest in **Virgo**—the Virgin. I've never been able to see a satisfactory figure of a virgin among these stars, but most viewers see the Y-shaped figure I've drawn. If you find her there, by all means draw her in! (But if you find something else there, draw that in. It's your book!)

Now you're looking nearly due south. Flanking Spica to the left or west is the kite-shaped figure of **Corvus**—the Crow. Farther west, almost to the westmost point on the horizon, lies the vast area populated mostly by dim constellations and even dimmer galaxies. To the east of Spica lies the dim, four-sided figure of **Libra**—the Scales. Libra almost looks like a bigger version of the kite-shaped Corvus, except its stars are dimmer.

To complete our tour, find Arcturus again. If you look at Boötes while facing east, its "wide-tie" figure is lying on its side with the point of the tie to the right. Just below the widest part of the tie, you'll find the lovely loop of stars known as **Corona Borealis**—the Northern Crown. This constellation is neither bright nor does it feature any interesting objects for small telescopes. However, it's pretty, so it stays in.

A bit farther to the right and closer to the horizon you'll find the **Keystone,** the central portion of the figure of **Hercules**—the Strongman. There are a number of stars around the Keystone that are supposed to represent various parts of Hercules' buffed-out body, but everyone I know only sees the Keystone.

Finally, if you look near the northwest horizon, you'll find a harbinger of the glories of the summer sky. That's Vega, a brilliant bluish-white star of magnitude 0.0. Only five other stars are brighter than Vega. It anchors an otherwise-ordinary constellation known as **Lyra**—the Lyre. The figure of this ancient stringed instrument is formed by a parallelogram (a squashed rectangle) of dim stars east of Vega.

Coma Berenices: A Galactic Cluster as Constellation

As I directed you to follow the paths to April's stars, I didn't have you stop at one of the more unorthodox of the 88 constellations, **Coma Berenices.** This group of stars is really not a classical constellation figure at all, but a huge, close galactic cluster, mixed with stars both closer to us and farther away than the actual distance of the cluster.

The effect is of a very big, somewhat dim "open" cluster. To the ancients, or at least to one desperate ancient, it suggested the hair of a minor goddess named Berenice, who in her earthly life was the queen to King Ptolemy III of Egypt. It was said that Berenice had lovely golden hair, which she had cut to celebrate the safe return of Old King Ptolemy from battle. A careless priest either failed to safeguard the hair, which was stored in the Temple of Aphrodite, or sold it to an unscrupulous wigmonger. Regardless, Berenice's hair disappeared. To save his skin, the quick-thinking priest convinced the royal couple her hair had been transformed into a constellation in the heavens.

You can find Coma by looking a bit north of the middle of a line between Arcturus and Leo's haunches. It's a nice object in binoculars. It's also part of the region of the sky I'll show you on April's ZOOM map. This area is sometimes called the "Realm of the Galaxies." I know you'll love it!

About the ZOOM Map for April

You'll remember from reading Chapter 9 that galaxies seem to congregate in clusters. One of the subjects of April's ZOOM map is the Virgo Cluster of galaxies, the richest such group visible in our skies. For a little variety, you can find one globular cluster and a fine double star.

VIRGO: "You can look at any kind of deep-sky object you want, as long as it's a galaxy!"

The French astronomer Messier cataloged deep-sky objects so he'd be able to avoid mistaking them for the comets he sought. If you look at April's ZOOM map, you might find it easy to believe that searching for a comet in Virgo gave Charles Messier a bad headache. I'll also remind you Messier did all

his observing with a telescope not nearly as good as the worst 7×50 binoculars on today's market.

You can *detect* the brighter galaxies in the Virgo Cluster with as little as a 3-inch telescope, although with galaxies the bigger the telescope you have, the better. To see all of the galaxies labeled on the ZOOM map, you'll need a 6-inch telescope and a dark sky. And, with galaxies especially, the more often you look at them the more you'll see. Also, since the light from most galaxies is very faint, you must be scrupulous in maintaining your dark adaptation. (You ARE reading this by a dim red light, aren't you?)

Messier plotted sixteen galaxies just in the center of the Virgo Cluster. There are many more, too many in fact to allow me to describe them all, so I'll give you some general information on the cluster, describe its brightest members, then suggest a "low stress" way of observing them.

The Virgo Cluster is about forty million light-years from us. There are, literally, thousands of

Map Symbols

Symbol	Magnitude
●	−1
●	0
●	+1
•	+2
•	+3
·	+4
·	+5 and +6

∞	Galaxy
×	Globular Cluster
○	Galactic (open) Cluster
◈	Nebula

galaxies in this cluster, including all those of our own Local Group, which lies near one edge. In the sky, the members of the Virgo Cluster are not just in the constellation Virgo. They're found as far north as Canes Vanatici, an invisible constellation just south of the handle of the Big Dipper, and as far south as Corvus, who crows about having some of these galaxies within his borders.

M64 is one of the brightest of the cluster, shining at a magnitude of 8.5. It's an unusual spiral, seen almost face-on. Unlike more classical spirals, M64 has a huge dust lane that looks like a crescent-shaped dark smudge in an otherwise fairly featureless oval-shaped object. M64 measures 3.5 × 7.5', so it's a fairly large object. To view it, start with your lowest power. Then bump up the power until you see the dust lane.

Another bright galaxy that also features a dust lane is the exquisite **NGC4565** in Coma Berenices.

It's a spiral like our own galaxy seen almost exactly edge-on. When viewed at low power in a 6-inch or larger telescope, it appears like a silvery needle. Higher powers (and steady seeing) reveal that the needle is split by a dark line—the dust lane. NGC4565 also has a pronounced central bulge.

If you rotated 4565 by 10° or so, you'd have a view similar to what you'll see when you view **M104**. Since you're viewing from slightly "above" the plane of its dust lane, M104's appearance is that of a big Mexican hat, defined by the slightly curved appearance of the dust lane, which is why it's called the "Sombrero Galaxy."

M100 is one of the brighter galaxies of this group, at magnitude 10.5. An almost face-on spiral, it takes a big telescope to see any hint of its detail.

M87 is a giant elliptical galaxy—one with no spiral structure and little gas or dust—which is one of the biggest of this type known. M87 has an enormous jet of hot gas being ejected from its center at very high velocity. Astronomers theorize that an enormous collapsed object, probably a really massive black hole, lies at the core of M87 and powers this jet.

You can find many of the outlying galaxies in the Virgo Cluster by carefully star-hopping to their plotted positions. However, you can use the tail star in Leo as the starting point to peruse the central group by allowing Earth's rotation to sweep your telescope across their positions in the sky. Here's how:

Let's assume you're using a telescope at 45× that shows a 1° field. Find the Lion's tail star and center it in your field. Now move your telescope slowly east until the first "fuzzy," **M98,** just appears at the edge of the field. If you have a clock drive on your telescope, turn it off. Now, just look as M98 slowly drifts across the field, only to be replaced by **M99** five minutes later and by **M88** ten minutes after that. Of course, there will be other dimmer galaxies slowly strolling through your field. (By the way, the name of the Lion's tail star is Denebola, which means "the Lion's tail star" in Arabic. Really!)

To pick up **M85,** start at Denebola, and move your telescope 3.5° north. For the telescope in the example above, you'd move it north three and a half times its field of view. Then move east to the first fuzzy, and let the view drift by the various objects until you come to M85, which will take about fifteen minutes. You can also continue your scan farther east to pick up April ZOOM's only globular cluster, **M53.** Remember, this tightly packed cluster is in *our* galaxy.

Using this method, you can scan across the central Virgo Cluster in an evening. Once you've become a bit familiar with where the brighter galaxies are, you can "zero-in" on each one and observe them at higher power. One word of caution: Watch out for the dreaded "galaxy overload"!

Exploring Leo's Galaxies, Plus One Fine Double

There are five bright galaxies just south of the figure of Leo—the Lion. **M65** and **M66** can easily be seen in the same low-power field. M66 is to the east and has a slightly irregular shape, although it is a spiral, and you can see the outer parts of its spiral arms. M65 is bigger and more conventional in appearance. On a clear, dark night, a 6-inch or larger telescope will show M65's dust lane at medium to high power.

M95 and **M105** are both fairly featureless galaxies of 10.1 and 10.5 magnitude, respectively. M95 is a spiral, while M105 is an elliptical galaxy. **M96** is a barred spiral whose bright bar can be seen in a 6-inch telescope on a dark night. The rest of its structure is only visible in something like my 20-inch.

One of the finest double stars in the sky can be found in Leo's sickle. It's Algeiba, the first star in the curved portion of the sickle. Various observers see this pair as yellow-yellow, yellow-green, yellow-red, and yellow-purple. I remember it as yellow-purple, but it's been a long time since I've observed a lot of doubles. Its components are of magnitude 2.1 and 3.4, and their separation is more than 4", and slowly getting wider. Of course, together, the two stars combine their light to look to the naked eye like a single star of magnitude 2.0.

A Final Word About Observing Galaxies

If all you want to do is detect some galaxies, a small telescope is adequate. However, you really need a 6-inch to see much detail in the brighter galaxies. A 10-inch to 12-inch really begins to show you a lot of detail, while a telescope of the 20- to 24-inch class turns many galaxies from "smudges-with-some-faint-detail" into objects that look just like—galaxies! (Going from the ridiculous to the sublime, I had the opportunity a few years ago to view a number of galaxies with the 33-inch telescope at the observatory of the National Autonomous University of Mexico on top of El Capitan, the 9200-foot mountain in Baja California. It was nearly a religious experience!)

About the ALL-Sky Map for May

May's sky gives us the last view of winter's constellations, as those of summer are higher in the sky each passing night. However, before we spend the summer looking at the gaudy attractions of the summer Milky Way, the more subtle delights of the "sky between the river of milk" beckons.

Arcturus and the Path West Along the Ecliptic

Just as Sirius dominates the winter skies, bright Arcturus dominates the late spring. In May, both the paths begin with this beautiful, −0.1-magnitude reddish-yellow star. To find Arcturus, look due south of the zenith. Only a bright planet could exceed the brilliance of Arcturus. Fortunately, the planets are restricted to following their own path, the ecliptic.

Before you "drop down" to the ecliptic, take a moment to trace the figure of **Boötes**—the Herdsman. This is a fairly large figure that looks to me like an upside-down "wide tie" from the seventies.

To get to the ecliptic and follow it west, drop down 30° toward the southern horizon to find a bright bluish-white star. This is Spica, the brightest star in the constellation **Virgo**—the Virgin. I can't find the figure of a "reclining virgin" here. But the Y-shaped group is what most observers see. Before finding Leo, take a quick look at the kite-shaped figure of **Corvus**—the Crow. It's a little south and mostly west of Spica.

Speaking of **Leo**—the Lion, you can find him by facing due west. Halfway from the horizon to the zenith, you'll see his figure formed by a sickle-shaped group for his head and mane and a triangle for his haunches. As figures in the sky are sometimes forced to do, our "king of the jungle" is diving head-first toward the western horizon.

Cancer—the Crab dimly occupies the next slot on the ecliptic. If you're learning May's sky with a pair of binoculars strapped around your neck, take a look about halfway between the top of Leo's mane

and the horizon. The huge galactic cluster you'll find there is M44—the "Beehive."

Just to the left of the Beehive lie Castor and Pollux, the "head stars" of **Gemini**—the Twins. Castor is the white star to the right, while yellow Pollux is to the left. Much past the northwest point on the horizon, Capella, the brightest star in **Auriga**—the Charioteer is just slipping over the horizon. With the departure of Gemini and Auriga, the last two of the winter constellations have gone off into the solar glare and then into the morning sky to work their way back into "prime time" next fall.

If you continue to your left from Capella, until you're facing due north, you come to the "squashed-W" asterism that's hugging the horizon. This is **Cassiopeia**—the Queen in Her Chair. Just above and to the right of Cassiopeia is her husband, **Cepheus**—the King of Ethiopia. Cepheus is a big, dim constellation, but he and Cassiopeia come as a set, so….

What the Dippers Are Doing

You'll find the end of the **Big Dipper**'s handle just north of the zenith. The rest of the Dipper is arcing down in the general direction of the northwestern horizon. Once you've traced the outline of the Big Dipper, you can use the two pointer stars at the end of its bowl to direct you to Polaris. Then, follow the arc of stars almost straight up from Polaris to the bowl of the **Little Dipper**.

Arcturus and the Path West Along the Ecliptic

Since you've just found the Big Dipper, you can use its handle to loop down to Arcturus once again, then you can continue in a grand loop to Spica.

Once at Spica, go back over to Virgo, Corvus, and Leo to reinforce what you've learned of their patterns and relative positions. Since it's taken you

ALL-Sky Map for May

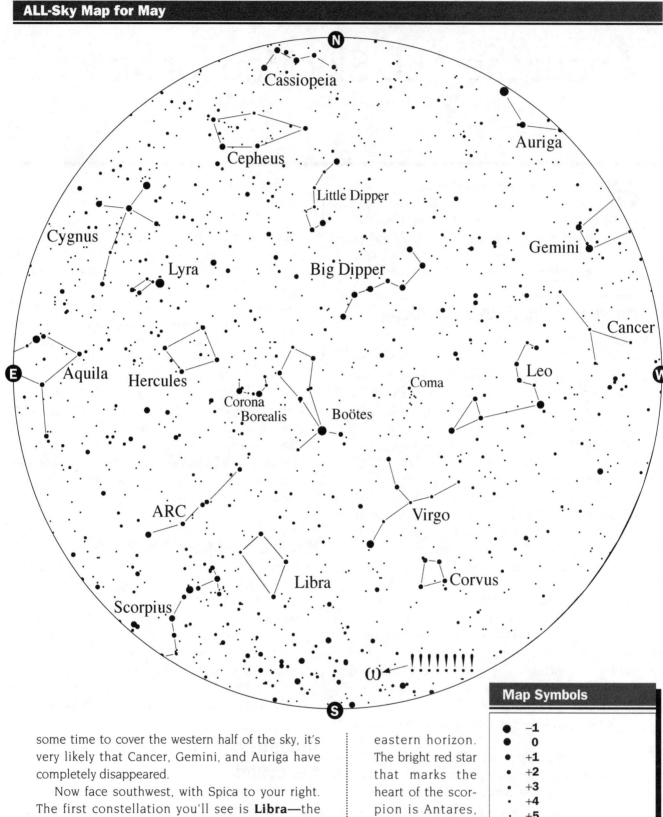

Map Symbols

●	−1
●	0
●	+1
•	+2
•	+3
·	+4
·	+5

some time to cover the western half of the sky, it's very likely that Cancer, Gemini, and Auriga have completely disappeared.

Now face southwest, with Spica to your right. The first constellation you'll see is **Libra**—the Scales. I don't see a scales here. Rather, I've drawn the rather common kite-shaped figure you see on the map.

Just below Libra, **Scorpius**—the Scorpion is poking its head and claws above the south-

eastern horizon. The bright red star that marks the heart of the scorpion is Antares, a magnitude 1.0 beauty. If you have a dark southeastern horizon, finish finding all the other constellation figures along the eastern horizon, then return to Scorpius to see all of its figure, including its magnificent tail. Sneak a look at June's map to see the figure.

Just to the left of Scorpius and Libra, you'll see an arc of stars labeled simply "ARC." That's what I and several friends see in that area, rather than the classical constellations of the region. You can read more about this asterism in June's commentary.

Due east, you'll see **Aquila**—the Eagle rising out of the gloom near the horizon. The eagle's eye is brilliant Altair, a magnitude 0.8 pure-white star. Altair and the two dimmer stars flanking it are one of the easiest asterisms to recognize in the summer sky.

Continue around the horizon until you're looking northeast. The bright-white star near the horizon is Deneb. It's either the top of the "Northern Cross," as I've drawn it on the map, or the tail of the Swan, as the (pre-Christian) ancients saw it.

Higher in the northeastern sky and to the right of Deneb is Vega, the brightest star in the constellation **Lyra**—the Lyre. The body of the Lyre, an ancient stringed instrument, is defined by the dim parallelogram of stars that extend below and to the right of Vega.

Between Vega and Boötes lie **Hercules**—the Strongman and **Corona Borealis**—the Northern Crown. Even though the ancients saw a full-blown (inflated?) strongman, the "Keystone" is what most people see, so I've drawn it here. Such are the vagaries of a constellation like the Northern Crown, a glittering half-circle of stars, placed right next to a common herdsman. If we're next to a herdsman, we must have returned to Boötes and Arcturus.

To make sure you'll remember all these constellations, why not go through them backwards, beginning with Boötes, then Corona, then Hercules, and so on.

An Off-ZOOM-Map Bonus—The Exquisite ω Centauri

The constellation **Centaurus**—the Centaur culminates these May nights, which means it crosses the meridian and reaches its highest point in its daily journey around the sky. Centaurus has a lot of bright stars, and its stars and those of Scorpius are part of a nearby cloud of stars called the Scorpius-Centaurus Association. Where this group fits into the spiral structure of the Milky Way galaxy depends on who you ask. However, it's relatively close to us.

In the temperate latitudes of the northern hemisphere, the southern horizon "cuts off" much of Centaurus, so I've not drawn its (rather fanciful) figure. If you don't live too far into the Frozen North, however, there is a bonus deep-sky object to be found far to the south in Centaurus these May nights. It's a magnificent globular cluster that's so bright, magnitude 4.3, that the early astronomical cartographers assigned it a star's name, even though it's clearly a very "fuzzy" star!

To find ω Centauri, look very near the southern horizon when Spica is exactly due south. This would ideally be maybe a half-hour earlier than the nominal time for which the May ALL-Sky map was drawn, or 09:50 P.M. To help out with the ID, I've drawn an omega (ω) at the spot on the May ALL-Sky map where Omega lies.

In any size telescope this globular is magnificent, partly because it's so close, only 17,000 light-years, but also because it's BIG. It's about 150 light-years across, which translates to half a degree on the sky. So, you're looking for a globular cluster that's the approximate angular size of the full Moon! Use low power to squeeze all of Omega into your field of view.

About the ZOOM Map for May

Because they're among the few constellations that many people learn in their youth, the Big and Little Dippers have their own little section in the commentary for each month's ALL-Sky map. But, in May, we dwell on the region around the North Celestial Pole.

Polaris is less than a degree from the pole, so it's used as a way to find north. Because of its unique location in the sky, this magnitude 2.0 star draws a lot of attention, despite the fact that there are 48 brighter stars in the sky. However, it's an interesting object all by itself, because it has a companion of the ninth magnitude at a distance of 18.5".

Bear Business

With apologies to Virgo, some of the best galaxies are to be found in or around the constellation **Ursa Major**—the Greater Bear, the official name for the group most of us know as the Big Dipper. Take another look at the ALL-Sky for May and you'll note that the sky between the Big Dipper and Leo seems to be occupied by a fair number of third- and fourth-magnitude stars that could be connected into constellations. And they were, of course, by the ancients.

The stars to the west of the bowl of the Big Dipper toward Auriga form the head, legs, and paws of Ursa Major. The stars of the Big Dipper make up the Bear's hindquarters and tail! South of the handle of the Big Dipper, we find two hunting dogs, possibly escaped from Orion's pack, a minor lion, and a fox, which probably lured the hunting dogs into escaping. No one sees these guys, so I haven't cluttered the maps with them.

Nor is the Big Dipper a universal figure. In the British Isles, it's known as the Plough or Plow. To escaped slaves traveling north along the underground railroad to freedom before the American Civil War, the Big Dipper was known as "the Drinkin' Gourd," possibly a holdover from their African heritage. What is universal about the Big Dipper is that it is host to some wonderful galaxies! Let's look at them.

The Big Dipper's Family of Deep-Sky Objects

Just as with the Virgo Cluster featured in April's ZOOM map, the galaxies near the Big Dipper are too numerous for me to describe all of them adequately. So I'll tell you about just four.

If you make a right angle with the last leg of the Dipper's handle and proceed along that line for 3.5°, you find **M51,** an exquisite face-on spiral galaxy that's on the edge of the Virgo Cluster. Since the last leg of the handle is 7° long, the distance to M51 is easy: It's half the length of that leg. This galaxy is actually located within the boundaries of **Canes Vanatici**—the Hunting Dogs. M51 is the object that puts Canes Vanatici on the map.

Since M51 shines at a total magnitude of 8.7, it can be found on a dark night with binoculars or a good finder. A small telescope shows a central concentration of nebulosity, surrounded by a weak haze. At the northern edge of the haze is another concentration of nebulosity.

On a still, dark night, a 6-inch reveals that the haze is not uniform, but has some structure. An 8-inch will show hints of the haze being spiral arms. When you move up to a 10-inch, the arms are fairly easy to see, and a 12-inch reveals M51 as the Whirlpool galaxy, a glorious spiral with two graceful arms. Larger telescopes reveal knots in the arms, and even a little detail in the intervening dust lanes. (Of course, as your skill increases, you won't need as much telescope to see the level of detail I've just described.)

This galaxy was the first to be recognized as spiral in shape, leading early astronomers to class many galaxies as "spiral nebulas." We now know that they are separate objects much farther away

ZOOM Map for May

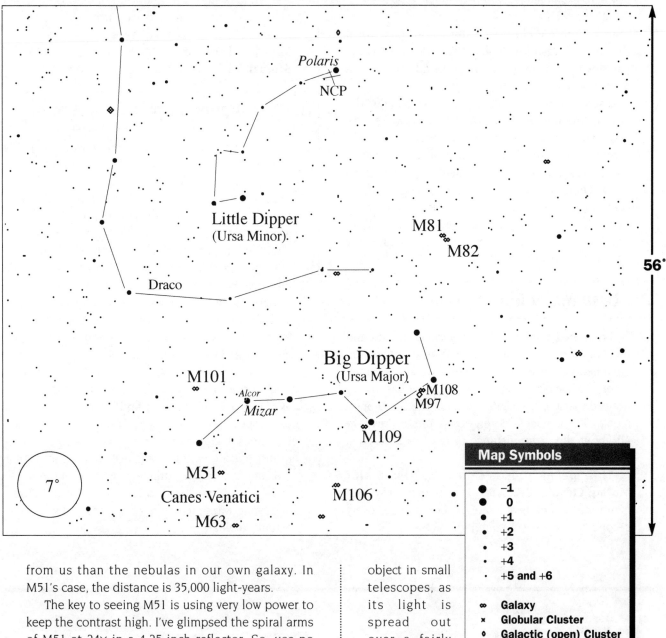

from us than the nebulas in our own galaxy. In M51's case, the distance is 35,000 light-years.

The key to seeing M51 is using very low power to keep the contrast high. I've glimpsed the spiral arms of M51 at 24× in a 4.25-inch reflector. So, use no more than 30× in a 6-inch scope, and as low a power as you can with a larger one. Once you're used to the view, bump the power up until you begin to see no improvement in the view.

What about the other concentration of nebulosity in M51? That's **NGC5195,** a dense irregular galaxy that's interacting with M51. "Interacting" is a polite word for "ripping off one of M51's spiral arms." In large telescopes, you can see that the end of one of the arms is stretched out right to NGC5195.

On the other side of the Dipper's handle lies **M101,** another face-on spiral. M101 is a so-so object in small telescopes, as its light is spread out over a fairly broad area. But it's big, almost one-third degree. Should you get a chance to see it in a 16–24-inch, don't hesitate to stand in line.

If you draw a line diagonally across the open end of the Dipper's bowl it points to a pair of galaxies, **M81** and **M82.** These two lie the same distance beyond the end of the bowl as the length of the diagonal itself. M81 is a big, bright spiral of magnitude 7.9, so it's easily seen in binoculars or your finder. Since M82 is only two-thirds degree away, both objects fit into the low-power field of telescopes. Use your lowest power to obtain enough contrast to see M81's spiral arms, then increase power to examine M82, a very irregular galaxy.

M81 is the flagship of a small cluster of galaxies that may be the closest cluster to our own Local Group. The cluster includes M82 and several other galaxies near them in the sky. Take a look within a degree or so and see what else you sweep up!

There's a nice planetary nebula just under the bowl of the Dipper. That's **M97,** also known as the Owl Nebula, because early visual observers thought it resembled an owl. M97 is one of the biggest of planetaries, those expanding shells of gas sloughed off by dying stars. Because it's so big, about 3 light-years across, and because it's being powered by what's left of its central star, the Owl is fairly faint. Use low to medium power on a very steady, very dark night to observe M97.

The Ursa Major Moving Cluster

From the Doppler shift of their spectral lines and their motion on the Celestial Sphere, astronomers are able to deduce the velocities of the nearer stars. When they began doing this on a regular basis, they noticed that the five central stars in the Big Dipper, along with perhaps a dozen other stars nearby, are all about the same distance from us and all have very similar velocities.

This sparse group is called the **Ursa Major Moving Cluster.** It's about 75 light-years away, and is moving toward us at a rate of about 9 miles every second. It'll be here in 1.5 million years.

Other studies show that Sirius, Denebola, the "tail star" of Leo, and about a hundred other stars also share a similar velocity with the Ursa Major Moving Cluster, indicating they may once have been associated. This group is called the **Ursa Major Stream.** We're pretty much in the middle of it.

The Big Dipper's Splendid Double

When you look at the Big Dipper, you'll probably notice that the star one-in from the end of the handle has a dimmer companion. The bright one is Mizar and its dimmer friend is Alcor. Although both are members of the Ursa Major Moving Cluster, they are not a true, gravitationally bound binary star. Their separation is about a fourth of a light-year. The ancients thought of this pair as a good test of a man's eyesight. However, they're separated by 12', or about a fifth of a degree. Mizar is magnitude 2.4, and Alcor is 4.0. They're not hard to separate. Either Alcor has brightened over the centuries, or the ancients had bad eyesight, or ????

What makes Mizar noteworthy is that it is a close double star in its own right, one of the finest in the sky. The pair are easy to split in the smallest of telescopes, being 14.4" apart. Their magnitudes are 2.4 and 4.0. Our friends, the spectroscopists, have found telltale shifts in each of their spectra that indicate each is an extremely close binary in itself, as is Alcor. So, when you look at Alcor and Mizar, and "split" Mizar, remember that each star you see is actually a binary.

About the ALL-Sky Map for June

June finds the constellations of summer well up in the eastern half of the sky as the last of the sparse spring constellations sink in the west. June is the first month that the summer Milky Way is well above the eastern horizon in the evening hours.

Saying Farewell to the Spring Stars

To make a quick loop along the western horizon, start with bright Arcturus, the brilliant reddish-yellow star just a bit south and west of the zenith.

ALL-Sky Map for June

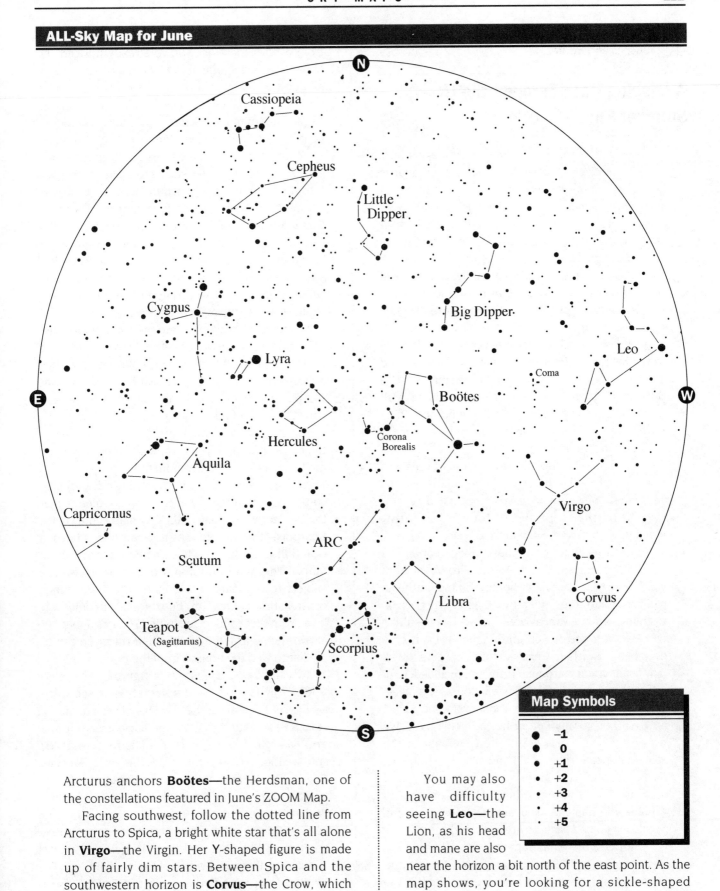

Map Symbols

●	−1
●	0
●	+1
•	+2
•	+3
•	+4
•	+5

Arcturus anchors **Boötes**—the Herdsman, one of the constellations featured in June's ZOOM Map.

Facing southwest, follow the dotted line from Arcturus to Spica, a bright white star that's all alone in **Virgo**—the Virgin. Her Y-shaped figure is made up of fairly dim stars. Between Spica and the southwestern horizon is **Corvus**—the Crow, which you may have difficulty seeing, owing to its proximity to the horizon.

You may also have difficulty seeing **Leo**—the Lion, as his head and mane are also near the horizon a bit north of the east point. As the map shows, you're looking for a sickle-shaped asterism. Leo's hindquarters, formed by the triangle you see on the map, are well above the horizon, as are the galaxies that lie in the region between his haunches and Boötes.

A Looping Path Through the (New) Summer Sky

Your path through the summer sky begins back at Arcturus and will cover the southeastern quarter of the sky. I think you'll be able to see this path best if you sit in a reclining lawn chair facing a little east of south. However, I'd hardly criticize a nice blanket on a south-facing slope, either. Don't forget the mosquito repellent.

Just east of Boötes and south of the zenith, you'll find a lively little half circle of mostly fourth- and fifth-magnitude stars. This is **Corona Borealis**—the Northern Crown.

Just east and a little north of Corona, you'll come to the Keystone, the asterism that's the central (and only recognizable) portion of **Hercules**—the Strongman. There must be a strongman hiding in all those stars. After all, the ancients found him. But I haven't, and neither do most other stargazers, so the Keystone will have to do.

You'll have little trouble locating bright Vega, the front star of **Lyra**—the Lyre. The remainder of this very small, very obsolete stringed instrument is formed by the trapezoid of fourth- and fifth-magnitude stars to the southeast of Vega.

So far, this looping path hasn't looped at all. It's been a straight shot from Arcturus to Vega. And, you'll have to look a bit farther toward the northeast to find Deneb, the tail star of **Cygnus**—the Swan. Cygnus is a fair swan, as you can see by consulting the ZOOM map for September. However, it's an even better Northern Cross, so that's how I've drawn it. Most amateur astronomers, myself included, refer schizophrenically to this constellation as Cygnus—the Northern Cross. You can, too, if you want.

We'll not go any farther north in this path. We now drop down to a point about 30° above the eastern horizon. There you'll find a bright white star, flanked on either side by stars two to three magnitudes dimmer. This is Altair, the eye of **Aquila**—the Eagle. As you can see from the figure, Aquila isn't a bad eagle as heavenly birds go.

Altair completes a "supergroup" known as the "Summer Triangle," the others being Deneb and Vega. You might want to take another look at the Summer Triangle, just to reinforce the patterns you've learned.

If you're learning the summer constellations from a rural location, you've already noticed the Milky Way meandering its way from a point just east of due north to a similar point on the southern horizon, just east of due south. As it passes through Cygnus, it splits into two recognizable "tributaries," continuing through the western part of Aquila.

Once it gets to the area just south of the tail of the Eagle, the Milky Way seems to merge once again into the beautiful, bright Scutum Star Cloud. Somewhere in this exceptionally rich knot of stars is **Scutum**—the Shield. Scutum's constellation figure begs the question: "If no one can recognize the shield as a shield, from what will the shield shield you?" No matter. The Scutum Star Cloud is a wonder, whether you view it with only the aid of your lawn chair, with binoculars, or a small telescope.

We're now traveling toward the heart of the Milky Way galaxy: **Sagittarius**—the Archer. There's a catch, of course. Where the ancients saw an archer, almost everyone now sees a Teapot. In fact, just as the nursery rhyme goes, this teapot is "short and squat." You can find the figure by looking low in the southeast just above the horizon. (Once you identify the easier Scorpion, you'll know whether or not you got the Teapot right on your first try.)

One of the most beautiful and graceful constellations in the sky is crossing the meridian at the time for which June's ALL-Sky map is drawn. It's **Scorpius**—the Scorpion. At its heart lies Antares, a beautiful reddish star whose name means "rival of Mars." The head and claws of the Scorpion are subject to the usual interpretation that surrounds constellation figures. Ah, but the tail! It could only be a scorpion's tail! To see it, find Antares about 30° above the southern horizon, then trace the tail to the left toward the horizon from there.

Just above Scorpius lies a figure I've labeled "ARC." This is what I and several friends see when we look at this part of the sky. Since I see an "arc" of stars, I've labeled it that. As the ancients saw it, the arc was the lower portion of three (related) constellations. From left to right, they were **Serpens Cauda, Ophiuchus,** and **Serpens Caput.** (In plainer language: The Tail of the Snake, the Snake Handler, and the Head of the Snake.) Serpens is one constellation with two separate territories, something like East and West Pakistan. It works about as well as the division of Pakistan did.

When I want to find an object in Ophiuchus or Serpens, I reference my "star-hop" to the ARC or one of the surrounding constellations. I'm not much interested in snakes. Nor do I see a snake handler in that part of the sky.

Below and to the right of the ARC, just past the head of the Scorpion lies **Libra**—the Scales. It's a constellation of the Zodiac, as are Scorpius and Sagittarius, but, otherwise, Libra has little to recommend it.

Moving quite a bit farther along the Zodiac returns us to Spica. A simple left turn returns us to Arcturus, where this path began. Then, you can follow the dotted line backwards to the handle of the Big Dipper for another episode of....

What the Dippers Are Doing

If you turn your recliner around, you can face north to take a look at the constellations that spend all their time near the North Celestial Pole. Look back over your left shoulder at Arcturus, then shift your gaze northward until you pick up the graceful curve of the handle of the **Big Dipper.** When you find the Dipper, you'll notice it's still in "dump mode" as it empties whatever liquid it's supposed to contain onto the northwestern horizon.

Use the bottom two stars in the Dipper's bowl to point down and to the left to Polaris the end of the handle of the **Little Dipper.** This figure is quite a bit harder than the Big Dipper, as its stars are dimmer. However, you should have little difficulty tracing the outline of the Little Dipper. Just remember that the two stars at the end of its bowl are considerably brighter than all the stars that lie between them and Polaris.

Just to the right of Polaris is the fairly dim figure of **Cepheus**—the King of Ethiopia, whose figure is relatively boring. The Milky Way running through his boundaries isn't particularly bright. Fortunately, the Milky Way in the two constellations on either side of Cepheus is considerably brighter.

Cassiopeia—the Queen in Her Chair lies below Cepheus. To the ancients, Cassiopeia's squashed-W figure represented Cepheus's Queen seated in her chair. Since the unsquashed portion of the W is usually represented as down, Cassiopeia is just beginning a six-month period during which she'll be seen upside down in the sky. Hardly a fitting position for an immortal queen!

Looking over your left shoulder, note that all you have to do is go look along the Milky Way past Cepheus to find Cygnus. Doing this will help you tie together the various paths.

About the ZOOM Map for June

June's ZOOM map provides a course in the comparative anatomy of globular clusters, those incredibly crowded groups of really old stars that seem to surround the Milky Way galaxy in a huge spherical shell.

A Strongman and a Strong Globular

Hercules—the Strongman has little to recommend him as a constellation figure. Aside from being the subject of numerous Grade B movies featuring various musclemen-turned-wanna-be-actors, Hercules is an innocuous group whose center, the Keystone, is the only recognizable portion of the constellation.

However, Hercules makes up for his lack of pizzazz as a figure by providing two of the finest deep-sky objects in the northern sky. The first of these lies two-thirds of the way along the western leg of the Keystone. Shining at magnitude 5.9, **M13** can often be detected with the unaided eye, especially if you use averted vision. However, don't expect to succeed with this trick unless you have perfectly dark skies. M13 is also known as the **Great Hercules Cluster.** It's one of the showpieces of the northern sky, and is sure to be high on anyone's list to show attendees at a summer star party.

ZOOM Map for June

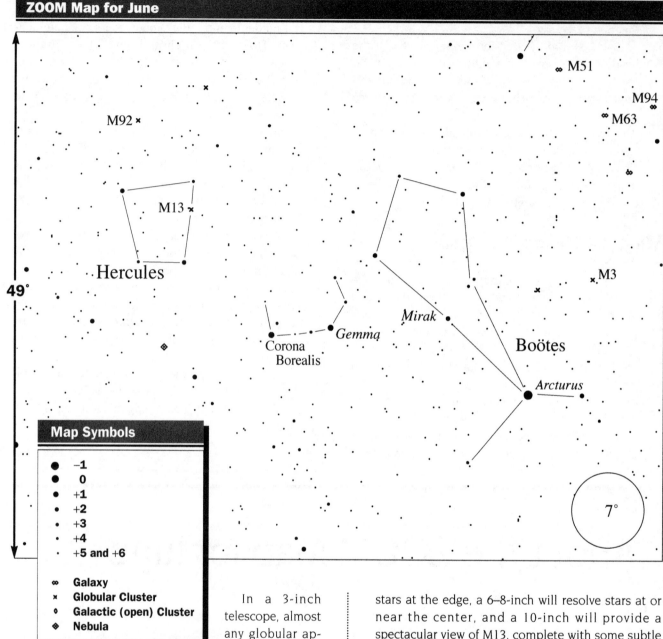

Map Symbols

Symbol	Magnitude
●	−1
●	0
●	+1
●	+2
●	+3
•	+4
·	+5 and +6

∞	Galaxy
×	Globular Cluster
◇	Galactic (open) Cluster
◈	Nebula

In a 3-inch telescope, almost any globular appears as a misty patch of light, bright at the center and dimmer toward its edges. That's because a 3-inch can't form the image of the dim individual stars in the typical globular that your eye can detect. It requires larger apertures (usually a 6-inch) to intercept enough light to make the individual stars detectable by your eye. Even a 6-inch will only resolve individual stars around the edges of a globular. If you can snag a look at a bright globular in a 10-inch telescope, you'll see stars resolved all the way across the center.

Because it's bigger and brighter than the average globular, the details in M13 can be seen in a smaller telescope than would normally be required to see them in a typical globular. So, a 4-inch will resolve

stars at the edge, a 6–8-inch will resolve stars at or near the center, and a 10-inch will provide a spectacular view of M13, complete with some subtle differences in color of the stars.

A large telescope will also intercept enough of M13's light that you can use high powers to probe deep into the core of the globular. There you'll find many beautiful arcs of stars. These arcs are merely phenomena of perspective and have no particular scientific significance. However, I know of no rule that requires that they be significant to be beautiful. So, enjoy them—again and again.

M13's apparent size of about a sixth degree, or 10', is a third the diameter of the full Moon. At its distance of 25,000 light-years, that makes the actual diameter of M13 about 100 light-years. Intensive studies of M13 indicate there are a million stars in this cluster. However, a little calculation reveals that the average density within the cluster is about one star per cubic light-year, maybe a little denser at the center. If the Sun were part of M13, our nearest star would be on the order of a light-year away, rather than the four light-years that is the case. But then we'd have a lot more bright stars in our night sky.

The brightest stars in M13 are red giants of about the eleventh magnitude. When we look at M13 in a very large telescope, these red giants aren't bright enough for our eyes to see them as red, because our eyes aren't very sensitive to red light. Instead, they appear slightly yellowish, compared to the dimmer stars in the cluster, which appear white.

M13 is the Babe Ruth of globulars, at least in Hercules. Just as few people can name the other 1927 Yankees besides the Babe, few stargazers pay much attention to **M92,** the other great globular cluster in Hercules. M92 is only a bit dimmer than M13, at magnitude 6.5, and only a bit smaller in apparent diameter, at about 8'. It's also farther away than M13, being 35,000 light-years distant.

Analysis of how the stars in globular clusters appear to have evolved leads astronomers to believe that globular clusters are among the oldest objects in our galaxy at 10 billion years. To put that in perspective, our Solar System was just forming when M13, M92, and their associates were already 5 billion years old!

Boötes' Claim to Fame—Arcturus + A Friend

Just as the life of a herdsman is usually peaceful but boring, the constellation of **Boötes**—the Herdsman is also mostly a bore. It has no "show-piece" deep-sky objects. Even M53, an interesting globular cluster at magnitude 8.7, is just over the border in the constellation Coma Berenices. How-ever, Boötes does have something to recommend it. There are two bright stellar objects of interest.

The first and most obvious is Arcturus, the fourth-brightest star in the sky and the brightest in the northern hemisphere. Shining at magnitude −0.1, Arcturus's name in Greek means "the Guardian of the Bear," an apparent reference to its position just off the tail of the Great Bear, ensuring that nothing can sneak up on him.

Arcturus is four times the mass of our Sun, but twenty-five times its diameter, putting out 115 times as much energy as the Sun. Since its area is so great, the surface temperature is a relatively cool 4200° Kelvin. Arcturus is cooler than a sunspot on our Sun; that's why it appears reddish-yellow to our eyes.

You can easily demonstrate the relative appearance of objects that emit "white" light. Take an ordinary household flashlight and shine it at close range on a piece of white paper at night. Try to remember the apparent color of the flashlight beam as it reflected off the paper. The following day, go outside in sunlight and shine the flashlight on the paper at the same distance you did the night before. You'd expect the beam to be as dim as it appears. After all, this is the Sun it's competing with. However, did you expect the beam to look so reddish-yellow? It's that way because the temperature of the filament in the bulb is well below the 5000° Kelvin of the Sun's photosphere. Remember, the cooler the source of light, the redder it looks to human eyes.

Arcturus is only 37 light-years away and is moving around our galaxy in an orbit that's steeply inclined to the plane of the galaxy. It appears as bright as it does because it's so close. In fact, Arcturus has only been visible to the unaided eye for maybe half a million years, and will only remain "naked-eye" visible for that much longer. Then it'll be lost in the galaxy, possibly forever.

You'll also find one of the finest double stars in the sky within the boundaries of "Boring Boötes." It's Mirak, the bright star "one up" from Arcturus on the east side of the figure of the Herdsman. Mirak shines with a total magnitude of about 2.4 as we see it with the unaided eye. However, it's actually a binary with a yellow-orange star of magnitude 2.5, and a bluish star of magnitude 5. Some observers describe the companion as greenish, but that's probably a trick of the eye. Although these two stars are definitely gravitationally bound, they've shown little change in position over the last 150 years, suggesting that the time it takes them to go around each other one time may be measured in thousands of years.

Mirak is separated by 2.9", so it's a tough double to split with a telescope smaller than a 3-inch. A 4-inch telescope should split this double easily. However, as always, if the atmosphere isn't steady, all bets are off.

A "Loose" Star?

Corona Borealis—the Northern Crown is a lovely little constellation that's devoid of any really interesting objects for the beginning observer.

However, its brightest star, Gemma, is thought to be a member of the Ursa Major Moving Cluster, that loose, very close galactic cluster that's made up of many of the bright stars in the Big Dipper. As you look at the Big Dipper, look all the way over to Corona. That's the true extent of what must qualify as the biggest deep-sky object in the sky, unless you count the Milky Way galaxy. (If you want to read more about this cluster, refer to the commentary on the May ZOOM map.)

About the ALL-Sky Map for July

For many sky-watchers, July is the first full month in which they can do any serious observing. The reason? June was the month when the kids finished school. Maybe the oldest graduated, or cousin Jenny got married one weekend, and you celebrated your anniversary the next. Or, possibly, June decided to be a repeat performance of April and May—lots of thunderstorms.

Whatever the reason, July seems to be the month when things slow down a bit, at least once you get past the Independence Day holiday in the U.S. As your reward for getting through winter and spring, July's skies are exquisite for the beginner just learning the constellations, for anyone just beginning to tour the sky with a small telescope, even for veteran observers. In fact, there's so much to see that these pages will only hit the high spots.

July's Path Around the Top of the Sky

Since it's nice and warm in the evening for most observers, you can stretch out on a reclining lawn chair to do your constellation study. First, face the western horizon. About 40° above the west point, you'll find reddish-yellow Arcturus, the brightest star in **Boötes**—the Herdsman, whose outline you can trace as he leans to the right on his squat legs.

To the left of the Herdsman near the horizon is Spica, the brightest star in the constellation **Virgo**—the Virgin. The haunches of **Leo**—the Lion are just slipping below the horizon to the north of due east. A large area of the sky that features no recognizable constellation figures lies between Leo's tail and Boötes. This area is uninteresting for constellation hunters, but contains many interesting galaxies.

Go back to Boötes. Just above the Herdsman's leaning figure, you'll find the lovely little half-circle of stars known as **Corona Borealis**—the Northern Crown. Just above Corona lies the group known as the Keystone, the central portion of **Hercules**—the Strongman. If you feel the need to see a guy with big muscles in the sky, I will invite you to invent your own figure. Almost everyone I know sees only the Keystone. (If you're having trouble finding the Keystone, remember it's very near the zenith.)

Back to Boötes, again. Possibly Boötes is leaning toward the **Big Dipper**, to his right in this view. Many legends claim Boötes is the Bear-Driver because he seems to drive **Ursa Major**, the Great Bear, around the sky. Confused? Don't be. The Great Bear is the constellation figure the ancients saw in Ursa Major. The Big Dipper is what most people see now.

Once you've found the Big Dipper, move your recliner so it's facing north. You can find the **Little Dipper** by using the end stars in the Big Dipper's bowl as pointers to Polaris, the end star in the Little Dipper's handle. Polaris is also the North Star, since it's less than a degree from the north celestial pole. The Little Dipper isn't likely to jump out at you, since only Polaris and the two stars at the end of its bowl are brighter than fifth magnitude. However, if you have a relatively dark sky, you should find it easy to trace the delicate arc of stars above and to the left of Polaris in this view.

To the right of the Little Dipper lies **Cepheus**—the King of Ethiopia. His five-sided figure is only slightly smaller than the Big Dipper, so don't be

ALL-Sky Map for July

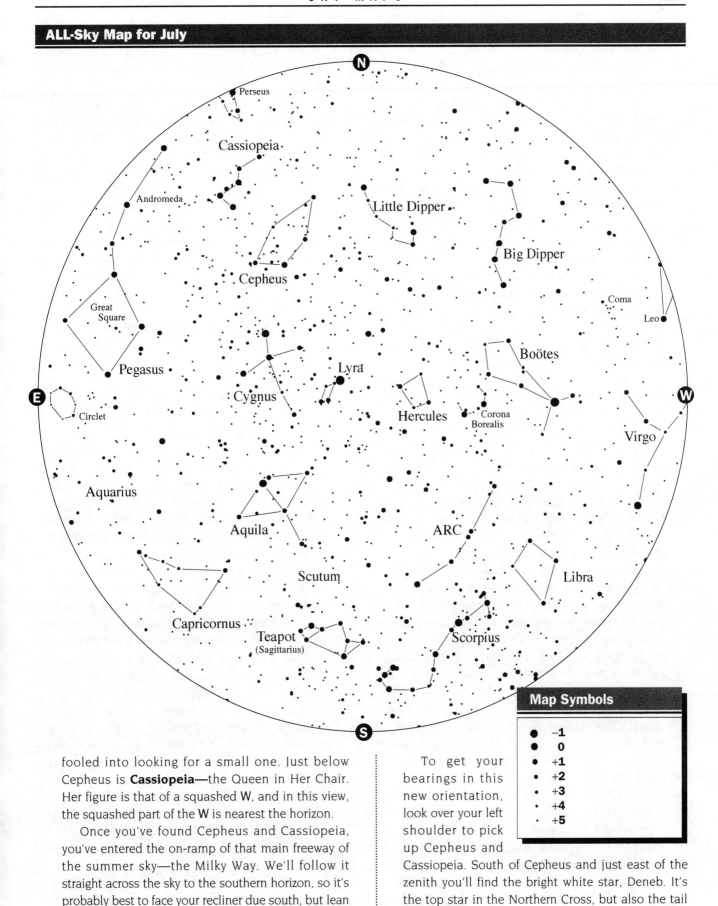

Map Symbols

●	−1
●	0
●	+1
●	+2
•	+3
•	+4
·	+5

fooled into looking for a small one. Just below Cepheus is **Cassiopeia**—the Queen in Her Chair. Her figure is that of a squashed W, and in this view, the squashed part of the W is nearest the horizon.

Once you've found Cepheus and Cassiopeia, you've entered the on-ramp of that main freeway of the summer sky—the Milky Way. We'll follow it straight across the sky to the southern horizon, so it's probably best to face your recliner due south, but lean it way back. A blanket on the lawn works fine, too.

To get your bearings in this new orientation, look over your left shoulder to pick up Cepheus and Cassiopeia. South of Cepheus and just east of the zenith you'll find the bright white star, Deneb. It's the top star in the Northern Cross, but also the tail star of **Cygnus**—the Swan. If you want to see the full outline of the Swan, check the commentary for the September ZOOM map.

Between Cygnus and the zenith blazes the blue-white Vega, the brightest star of **Lyra**—the Lyre. I'd be a lyre, too, if I told you that the rest of Lyra is very bright. It isn't. This stringed instrument, which various godlets played in ancient Greece, is made up of Vega and a little parallelogram just below it and to the left.

Below Cygnus you'll find the figure of **Aquila**—the Eagle. The eye of the Eagle is the white star Altair, which you can find easily by noticing that it's flanked on either side by stars of the third and fourth magnitude.

Aquila's tail melts into the Scutum Star Cloud, part of the constellation **Scutum**—the Shield. I can't see any recognizable figure here, much less a shield, so I've not drawn one to confuse you (or me). If you're studying the constellations under a dark sky, the cloud should be what attracts you to Scutum.

You're ready now to drop down near the southern horizon to learn the constellations of the richest regions of the sky we residents of the northern temperate zone can see. Start with the figure most people see as the Teapot. Its formal name is **Sagittarius**—the Archer, but I'm betting you see a teapot, not an archer.

The center of our Milky Way galaxy lies north of the spout of the Teapot. In fact, the "steam" coming out of the spout is a star cloud aligned almost exactly with the galactic center.

To the Teapot's right, you'll have the pleasure of tracing the outline of **Scorpius**—the Scorpion. The figure of the scorpion's tail is, arguably, the most graceful in the heavens. It starts below and to the right of the Teapot with a nice bright pair of stars, then loops gracefully down, around, and back up to the heart of the Scorpion, Antares. Ares was the ancient Latin name for Mars, the Red Planet. This bright reddish star is so named because it was thought the "rival of Mars." The figure of Scorpius ends with the graceful arc of stars just beyond Antares. Others have drawn pincers beyond that point. If you see them, by all means, add them!

Just to the right of the Scorpion lies **Libra**—the Scales. This constellation is as boring as Scorpius is interesting. It is, however, one of the constellations of the Zodiac, so it gets some famous visitors, such as the Sun, Moon, and planets.

Just above Scorpius lies the arc of stars I've drawn as "ARC." This isn't a recognized constellation, it's an asterism I see when I look just above

Scorpius. It's seen by others, too, but is by no means a universal figure. (For that reason, I hesitated to call it Rick's ARC. I also feared my fellow amateur astronomers might stone me!)

The ARC is really part of **Serpens Cauda, Ophiuchus,** and **Serpens Caput,** the Tail of the Serpent, the Serpent Handler, and the Head of the Serpent, from left to right. You can see many interesting objects in this region, but I doubt that one of them is a big guy holding a snake.

The ARC loops back up to Arcturus, which you can see over your right shoulder, due west. We've completed a long path around the sky and there are only a few constellations along the eastern horizon to go, then you've completely toured July's sky.

A Sea-Goat and a Huge Dipper

To reinforce some of what you've just learned, loop backwards through ARC, Libra, Scorpius, and Teapot. Once you've found these figures again, look to the left, or east of Teapot to find the big, fairly dim figure of **Capricornus**—the Sea-Goat. Even though it's not bright, I stubbornly see this figure, even though I wasn't born under the sign of the goat.

However, I don't see this figure as a goat. When I was a kid, I saw it as a seal, balancing an imaginary ball on its nose. If it was holding a fish in its mouth, the Eagle would probably swoop down and steal it!

If you face east these July evenings, you'll be able to see the figure of an enormous square just to the left or north of due east. This figure is the Great Square of Pegasus, really part of a constellation formally known as **Pegasus**—the Winged Horse. Beware the drawing of the Square this month. It's found in the region of the map that distorts figures the most. It's really a better square than you see on the map.

Attached to the Square is a graceful arc of third- and fourth-magnitude stars that make up **Andromeda**—the Chained Lady. The corner of the Square to which the arc is attached is really within the boundaries of Andromeda. However, no one sees either a chained lady or a winged horse. What many see is either the Great Square or a great dipper, as I've drawn it. As the month wears on, this great dipper rises higher, and you can see it better.

And the Chained Lady? You can read all about her unfortunate plight in the commentaries for the fall constellations.

About the ZOOM Map for July

This commentary is the hardest to write because there are so many wonderful objects to describe and only so much space in which to do it!

A Wealth of Globulars for Your Viewing Pleasure

July's ZOOM map plots two dozen globular clusters, a large fraction of the known total in our galaxy. This commentary includes descriptions of M22 and M4. However, please don't fail to observe the other eleven globulars I haven't the space to describe. They're all worth a look.

M4 is probably the easiest globular cluster in the sky to find. It's a little over a degree west of Antares, so all you have to do is center your telescope's low-power field on Antares and slowly slew it west until you see M4 enter your field of view.

M4 is one of the nearest globulars to us at a distance of something like 6000 light-years. It's also one of the biggest of the globulars, being about 20', or a third of a degree in diameter. As globulars go, M4 is a bit on the loose side, mostly because there are few giant stars to "fill up" the central part of the cluster as there are with many other globulars. That there is a lot of dark material between us and M4 also contributes to its loose appearance. If this material weren't there, M4 would shine almost a magnitude brighter than its actual brightness of 7.

Because its stars are a bit dimmer than the typical globular of its size and distance, M4 really doesn't begin to look like a globular until you look at it through a 6-inch or greater aperture. However, a good 4-inch will show some stars around its edge on a steady, dark night.

Were it not so far south, **M22** would be the "M13" for northern observers; it's that good. You can find this sixth-magnitude object just a bit north and a few degrees west of the top star in the Teapot. It appears bigger than M13, at about 17' in diameter,

and it's closer at about 10,000 light-years, but M22 appears to have "only" half a million stars, where M13 has a million or so. M22 also has more material absorbing its light than M13 does.

Regardless, M22 is one of the showpieces for northern observers. It is easily seen as a globular in a 4-inch and is a glorious sight in anything greater than a 6-inch. Although you should always begin with low power, M22 can stand almost the highest power your seeing will allow.

Bright Open Clusters—a Feast for the Binocular Viewer

On a dark night, you can't point your binoculars to any point in the field of July's ZOOM map and NOT find a galactic or open cluster of stars. They're all over the place. I'll describe M11 and M26 in my commentary on the ZOOM map for August. In this space, I'll describe M6, M7, and M24. Although I'll not be describing them, the others found on July's ZOOM map are worth finding.

If you have a dark sky and a low southern horizon, you'll have little trouble finding **M6** and **M7.** They're easily visible to the naked eye, just off the tail of the Scorpion.

Of these two clusters, M6 is the northernmost. It's located just out of the Sagittarius Star Cloud, so even though it's the dimmer of the two at magnitude 4.2, it seems to stand out just as well. M6 is often called the "butterfly cluster," because its brighter stars form a mini-constellation that is alleged to resemble a butterfly. I've not seen that figure, but you might.

M6 is possibly 1200 light-years away, making it one of the closer galactic clusters. It's about 10 light-years across, and contains sixty to eighty stars, but forty of them are as bright as the tenth magnitude, so it looks bright in almost any instrument. At an age of about 100 million years, M6 is quite young by cluster standards.

ALL-Sky Map for July

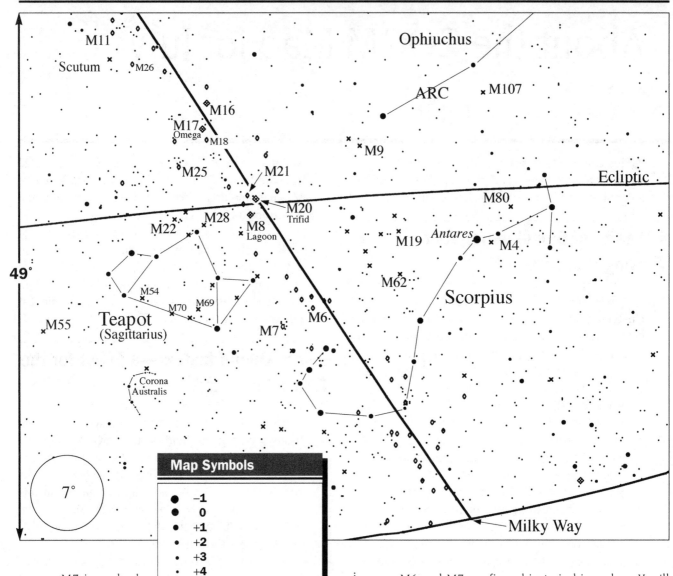

M7 is embedded in the Sagittarius Star Cloud, so its eighty stars are backed up by the "cloud" of much dimmer stars that form the Milky Way. It shines at magnitude 3.3, so it's easily seen with the naked eye, sometimes from suburbia as well as from the boonies. M7 is about 800 light-years distant. It's almost a degree in apparent diameter, which translates to about 13 light-years in diameter.

Of its approximately eighty stars, all are brighter than the tenth magnitude, accounting for the high brightness of the cluster as a whole. Studies of how these stars have evolved indicate that M7 is perhaps 250 million years old.

M6 and M7 are fine objects in binoculars. You'll get a good view if you pick a site with a very low southern horizon whose skies are polluted with little extraneous light. To view these far southern objects, try leaning your elbows on the roof or hood of a car or pickup, using them as the legs of a "human tripod." (The other "leg" of the tripod is the bridge of your nose.) If you view these two objects in a telescope, use the very lowest power you have; these guys are big!

There's some doubt as to what Messier saw when he added M24 to his famous list. Most people think it's the Small Sagittarius Star Cloud, a major constituent of the "steam" coming out of the Teapot's spout. You're not going to see this object well in anything but a telescope of very low power, or through a good pair of binoculars.

A Trio of Bright Nebulas

Three of the finest nebulas visible to northern observers are found in the center of July's ZOOM map. They're M8 and M20 to the south and M17 to the north.

Eight degrees north and a degree west of the top star in the Teapot you'll find **M17,** the incomparable Omega Nebula. This object is an emission nebula that features very high surface brightness, so its brightest parts are easily visible even in a small telescope.

The Omega got its name because the nebula looks somewhat like the Greek letter omega. However, I like its other common name, the Swan Nebula. Once you've seen the Swan, I suspect you'll like that moniker better, too.

The Swan is big, only a bit smaller than the full Moon, so you needn't use high power both to see detail in the nebula and to keep contrast high. (I rarely use more than 100× when viewing this object with my 20-inch reflector.) Since the Swan is about 6000 light-years distant, its angular size translates into a diameter of about forty light-years. The bright central bar is about a dozen light-years across. The Swan is really comparable in size and brightness to the Orion Nebula. Unfortunately, it's four times farther away. Still, it's the second-finest emission nebula visible from our northern skies.

On the other "side" of the Small Sagittarius Star Cloud we find two more bright nebulas. **M8,** known as the Lagoon Nebula, is 5° north of the end star in the Teapot's spout. It's a fifth-magnitude emission nebula fully a half by one degree in extent. Although its total brightness is greater than the Swan, its light is spread out over six times the area, so its surface brightness is not as great as the Swan. M8 got its name because it has a dark nebula crossing it mostly from east to west. This dark nebula is the "lagoon."

Embedded within the Lagoon Nebula is a galactic cluster of stars that formed within this cosmic nursery. Many more are certain to form, as the dark globules out of which stars are thought to form can be seen on long-exposure photographs of the Lagoon. This nebula is about five thousand light-years away and is about sixty by one hundred-twenty light-years in total extent.

Because it's so big, you'll be able to see this nebula in binoculars. You should even be able to glimpse the "lagoon" under dark skies. In a small telescope, you can use medium power to resolve some of the detail within the nebula. A sufficiently large telescope may not allow you to see all of the Lagoon Nebula, pointing out that for some objects in the sky a big telescope is not necessarily a better one.

M20 may be found just 1.5° almost due north of the Lagoon. This nebula is a smaller version of the Lagoon, in that it has streaks of dark nebulosity dividing it into sections. Since we see it in three almost-equal sections, M20 is popularly called the Trifid Nebula.

The Trifid is smaller than the Lagoon, being only about a quarter degree across. At its distance of possibly 6500 light-years, that angular size translates into a diameter of thirty light-years. There's a triple star almost in the center of the Trifid. It's brightest member appears to be the star that provides all the ultraviolet light that excites the nebula into emission in the visual region of the spectrum.

You may not be able to detect the divisions in the Trifid with a small telescope, as they're not large features. A 6-inch used on a dark night will show them, but it takes an 8–10-inch to show them plainly. In a large telescope, this nebula is a truly glorious object.

Exploring for Dark Nebulas

For good reason, books written for beginners almost never mention how to find dark nebulas. That's because they require dark skies and perseverance to be seen. My theory is that the worst that'll happen is you won't see them, at least not on the first try. But they're among the most beautiful objects in the heavens, even if their beauty is both subtle and difficult to see.

Dark nebulas abound in the region just west of the bright objects we see in the steam coming from the Teapot. That's because we're looking under the dust lanes that girdle our galaxy by virtue of our position out of the galactic plane.

The best of the dark nebulas in this region is **Barnard 72,** an S-shaped patch of "no-stars" that's just about 9° west of the Lagoon. To find B72, you must have very dark skies, no moon, very low power, and your eyes must be completely adapted to the dark. 7×50 or 10×70 binoculars will show this object well, but a 4-inch reflector operating at 16–20× seems to show it best. If you succeed in finding B72, look around the area for more. They're there.

About the ALL-Sky Map for August

July gave you the best view of the southern Milky Way, but August is the month when the summer Milky Way stretches straight across the sky, dividing the heavenly vault in half. This is the month when you can easily while away an entire evening as you tour up and down the Milky Way with binoculars. Many of the objects you'll see in binoculars are described in the commentary on the August ZOOM map. This map covers the constellations of Aquila, Capricornus, and Scutum, as well as some bit players that aren't found on the ALL-Sky map. First, a tour of all of August's sky.

The Western Path, or The Herdsman Takes His Leave

In August we say farewell to Arcturus, the star that's anchored our paths since late last spring. To find Arcturus, face due west, and settle into a reclining lawn chair. You're looking for the bright reddish-yellow star just above the horizon a bit north of due west. Arcturus anchors **Boötes**—the Herdsman, who's almost standing on the western horizon.

We'll look at the Big Dipper and friends a bit later. For now, look just to the left of Boötes. The little semicircle of stars you see next to his shoulder is **Corona Borealis**—the Northern Crown. Above the Crown, halfway from the horizon due west to the zenith, you'll find a "squashed" square of four stars. That's the Keystone, central portion of **Hercules**—the Strongman. Various versions of the full figure of Hercules exist, but all seem to put his head toward the north and his feet to the south. If you see a strongman here, add to the Keystone figure I've drawn.

Far to the left of Boötes stretches the ARC, an amalgamation of the bottom row of stars that make up **Serpens Cauda, Ophiuchus,** and **Serpens Caput.** In this view, these three constellations occupy the area immediately above the ARC and to the left of Boötes/Corona/Hercules. They're meant to represent, from left to right, the Tail of the Serpent, the Serpent Handler, and the Head of the Serpent. I haven't drawn these three constellation figures for two reasons: First, I don't see them, and second, there's just not much demand for "trained-snake-acts" in the celestial circus anymore, so almost no one has any idea what a snake-handler looks like! (If you think such acts are poised to make a comeback, draw these guys into your maps.)

The ARC parallels the horizon to the left of due west. Directly below it, dim **Libra**—the Scales sinks into the gloom. If you have a dark sky and a low southwestern horizon, you might see Libra, but the important reason for being able to see down to the horizon is to see Scorpius sprawled out along it.

Scorpius—the Scorpion is one of the most graceful and easily recognized constellations in the sky. Although its telescopic objects are easiest seen earlier in the summer, I've always enjoyed seeing Scorpius in this attitude. Note, however, that the map projection has stretched the scorpion a bit.

Just above the scorpion's tail lies the Teapot, which is the recognizable figure for **Sagittarius**—the Archer. Just as the ancients saw the archer shooting an arrow into the heart of the scorpion, we "moderns" see the Teapot preparing to pour scalding water on his tail!

You've probably been sliding your gaze (and your recliner) to your left as you follow the setting constellations around the horizon from west to south. You can now face due south and lean your recliner back some for a tour of the Milky Way and its constellations.

You've likely noticed the Milky Way isn't a smooth "river of light" across the sky. Rather, it's a collection of clouds of stars. The individual stars in the clouds aren't visible to your eye because they're too dim. However, their collective light you see as the clouds.

The Sagittarius Star Cloud is just above the spout of the Teapot. A bit farther along the Milky Way, forming the "second puff of steam" is the Small

ALL-Sky Map for August

Map Symbols

●	−1
●	0
●	+1
•	+2
•	+3
•	+4
·	+5

Sagittarius Star Cloud. Well above the Teapot, you'll find the Scutum Star Cloud. It looks like there's a breeze blowing from right to left across the sky, carrying steam from the Teapot up and across the "top of the sky."

The Scutum Star Cloud lies in the constellation **Scutum**—the Shield. The Scutum Star Cloud is

shielding you from seeing the Shield. It's a really weak figure, so I've not drawn it.

Just to the left of Scutum, take a detour off the Milky Way to trace the outline of **Capricornus**—the Sea-Goat. Others draw a much more involved figure for Capricornus, probably feeling obliged to do so because it's one of the constellations of the Zodiac. I have neither the interest nor the imagination for such figures, so I've drawn it as you see it. To me, it looks like a sea-lion, not a sea-goat.

Above Scutum and Capricornus you'll find the figure of **Aquila**—the Eagle as he patrols the Milky Way as it crosses the celestial equator. Aquila is easy to see, especially his bright eye, the brilliant white star Altair, flanked by a pair of dimmer ones.

Depending on your latitude, the figure of **Cygnus**—the Swan is either at or just south of the zenith in August. Since I suffer from a rare form of schizophrenia seen only in people who doodle maps of the sky, I describe this constellation as a swan, but I've drawn it as the more familiar Northern Cross. Actually, I like the sound of the word Cygnus, but see the shape of the cross in the sky. (September's ZOOM map shows the whole swan.)

Just to the right or west of Cygnus, you'll find the brilliant blue-white star Vega and its delicate little parallelogram of stars that form **Lyra**—the Lyre. Vega, Deneb, and Aquila form a supergroup of stars known as the Summer Triangle.

The Milky Way at "The Top of the Sky"

The Milky Way splits at Aquila and a dark rift begins to the west of Altair, extending along our galaxy toward Cygnus. It runs through Cygnus, paralleling the long arm of the Northern Cross. The **Cygnus Star Cloud** extends along the longer "vertical" arm of the cross past where the arms intersect. It's one of the most beautiful areas of the Milky Way galaxy, although it doesn't feature the bright showpiece objects that are found in the Clouds of Sagittarius or Scutum.

Note that describing these objects as "clouds" reflects our bias toward bright objects being regarded as "real," and dark objects as lacking substance. In actual fact, the various clouds of stars we see are portions of the spiral arms of our galaxy we see through holes in the dust lanes between the arms. The Incas, those hardy natives of the Andes, got so vivid a view of the sky from their homes more than two miles high that they developed two systems of constellations. One is like our Eurocentric "connect-the-dots" system. The other involves their perception of the shapes defined by the interplay of the dark and bright clouds of the Milky Way galaxy!

What the Dippers Are Doing

July's path was a continuous one that took you all the way around the sky. That allowed "What the Dippers Are Doing" to take a one-month vacation.

This month's path gets confusing without splitting it up, so "WtDAD" is back. To find out "WtDAD," turn your recliner due north.

About 30° to the left, or west of due north, you'll find the **Big Dipper,** finally in position to keep some of its contents from splattering the northern horizon. Follow the end stars in the Dipper's bowl to Polaris, the North Star. You can then trace the delicate arc of the handle of the **Little Dipper** up and to the left of Polaris. In this view, the Little Dipper holds no liquid.

Above and to the right of Polaris you'll find the five-sided figure of **Cepheus**—the King of Ethiopia. He's sufficiently boring a constellation that he's standing on his head to try to attract your attention. Ignore him, and go to his right to see a figure that looks like a squashed 3 in this view. It's Cepheus's queen, **Cassiopeia,** seated on her throne. Like Cepheus, she's suffering the ignominy of being upside-down.

Once you find your way to the royal, upside-down couple, you're back in the Milky Way. Above Cepheus you'll find your way back to Cygnus, tying your trip along the first path of August to this section. Below and to the right of Cassiopeia, you'll find **Perseus**—the Hero, as he climbs out of the gloom on the northeastern horizon.

The Path on the Home Stretch

Now face your recliner due east to trace the constellations of fall that are just rising. About halfway between the zenith and the horizon you'll come to a group of stars that form a huge square standing on one of its corners. It's the **Great Square of Pegasus,** which the ancients saw as a winged horse.

In order to draw the Great Square, I've stolen a star from **Andromeda**—the Chained Lady. She's the graceful arc of stars that flows from the left star of the Square as we see it in this view. (Some observers see two arcs.) I don't see a chained lady in Andromeda's figure, nor am I particularly fond of chains (or whips). So, I've drawn Andromeda and Pegasus as most people see them today, as a square with a handle. Some stargazers I know think of this group as the Great Dipper.

Below Andromeda lie the thoroughly unrecognizable **Aries**—the Ram, and **Pisces**—the Fishes. To avoid failing to draw the figures of the many constellations of the Zodiac, I've drawn the outline of a nice little group called the **Circlet,** which is part of Pisces. Although the Circlet isn't bright, it is easily found by most observers, so I put it in. To the right of the Circlet lies **Aquarius**—the Water Bearer. Aquarius is my astrological sign, yet I am neither a flower child nor can I see the outline of the Water Bearer. (Possibly my birth certificate is a forgery.) I have drawn the smaller Water Jar, though.

To the right of Aquarius is Capricornus, which brings you back to known territory. To reinforce the patterns you've learned, why not start at Capricornus and follow the constellations along the Milky Way to the northeastern horizon, and then along the horizon back to Capricornus. (Just like chocolate, you really can't get enough of the Milky Way!)

About the ZOOM Map for August

Plotted on the ZOOM map for August are six globular clusters, three galactic clusters, and one exquisite planetary nebula. I'll also direct you to several features within the Milky Way that aren't plotted. Although this region of the sky isn't as showy as the area near the galactic center that I featured on July's ZOOM map, I doubt if you'll be disappointed by what you'll see.

From the Scutum Star Cloud to the Altair Rift

The **Scutum Star Cloud** is worth a long look in binoculars. This intense cloud of faint stars appears to the unaided eye as a dense bright milky cloud. Binoculars reveal it's made up of thousands of dim stars, but not all its stars are resolved. A telescope used at low power reveals that the Scutum Star Cloud is made up of uncountable dim stars.

My favorite view of the Scutum Cloud is through binoculars, but you can also view it at low power by scanning slowly across it, using overlapping scans. Buried deep within the Scutum Star Cloud lies a true treasure. Just south and west of the tail of Aquila—the Eagle in the northern part of the cloud lies **M11,** one of the finest, richest galactic clusters in the sky.

Your first view of M11 will probably be in binoculars or at low power in a small telescope. M11 appears to be a bright globular cluster in small instruments. A 4-inch will resolve many of M11's stars, while a 6-inch is needed to see that it's a tightly packed galactic cluster, not a globular. Since its brightest stars are eleventh-magnitude giants, M11 will stand a lot of magnification, so you can break the rule that galactic clusters are best viewed only at low power. Treat M11 like the globular even if it isn't.

Since it's a compact 10' in size, M11 isn't too big to make viewing it in a larger telescope unrewarding. I've often gazed at M11 in my 20-inch. It's always a source of oohs and aahs, as those white giants just seem to jump out of the eyepiece at me.

M11 lies at a distance of about 5500 light-years and is about 15 light-years across. Because it's so compact, it resembles a globular cluster. The key to its "galactic clustertude" is that its age, half a million years, is a small fraction of the age of a typical globular. Globulars are old, some as ancient as ten billion years.

About 3.5° west and (mostly) south of M11 lies **M26,** another galactic cluster. Were it not for its gaudy neighbor, M26 would be better known. It's a ninth-magnitude, compact cluster of twenty to thirty stars whose distance is thought to be a bit closer than M11's. Both clusters are closer to us than the Scutum Star Cloud.

If you use your binoculars to scan across the body of the Eagle, you'll notice that the number of stars seems to fall off as you scan northwest of Altair. You've found the **Altair Rift,** an area of the dust lane that obscures our view of the Sagittarius Arm of our galaxy beyond. You can follow this rift along the Milky Way into Cygnus.

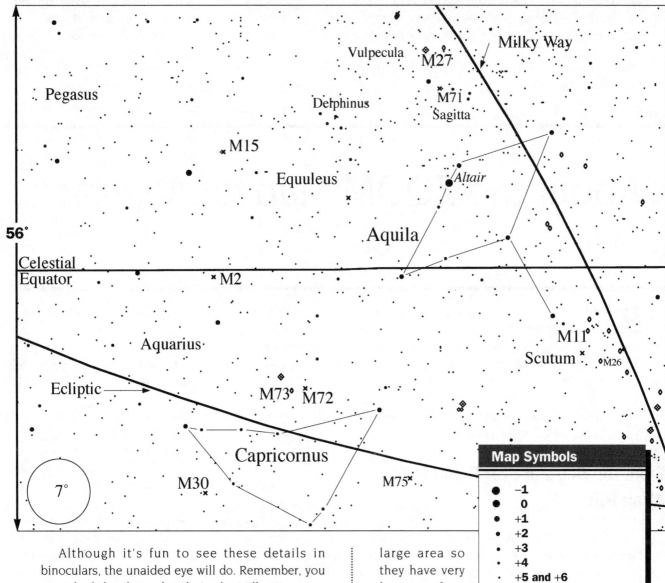

Although it's fun to see these details in binoculars, the unaided eye will do. Remember, you must look for these details in the Milky Way on a clear, very dark, moonless night. Give your eyes at least a half hour to adapt to the dark before trying to see these dust lanes in the Milky Way. An hour is even better.

THE Planetary

Because they're powered by the star that just created them, planetary nebulas can't have extremely high intrinsic brightness. There's also the size bugaboo. If they're new and therefore fairly bright, they've not had time enough to expand to a size that makes them easy to observe at relatively low power. If they're old and therefore fairly big, all the power they radiate is spread over their

large area so they have very low surface brightness.

The solution, of course, is to arrange to be very near a planetary nebula. In the case of **M27,** in the constellation **Vulpecula**—the Little Fox, we're only some 500–900 light-years away. The "Dumbbell Nebula," as it's called, is the biggest and brightest of all the planetary nebulas visible in the sky. It shines at magnitude 7.3, and can easily be seen in binoculars or a good finder. Its size, 5×8', is sufficient that it takes little magnification to make it large enough to see detail without washing it out.

The name comes from its appearance in a small telescope. To many early observers, M27 looked like a dumbbell, with two roughly spherical concentrations of nebulosity separated by a thinner bar. Later generations of observers have seen an hourglass shape. However, I suspect all of us have been influenced by photographs showing that shape.

On a dark, steady night, a 3-inch telescope will show the Dumbbell as a slightly squared-off oval object at medium power. A 4-inch will begin to show the distinctive "dumbbell" or "hourglass" shape. A 6-inch allows you to see a slight bumpiness in the nebulosity, while an 8–10-inch shows even more detail. Larger telescopes allow a glimpse of subtle red and purple tints in some of the brighter parts of the nebula. (Kids will see this more often than we older folks. Their eyes have much greater sensitivity at the extremes of the visual spectrum than adults' eyes.)

A Slew of Globulars

All of the six globular clusters labeled on the ZOOM map for August are worth a look, especially if you're using a telescope of 6-inch aperture or larger.

Of this group, **M15** is perhaps the nicest, as it's both relatively bright at magnitude 6.5 and quite compact. Since it's so compact, M15 has always allowed a higher-power view than some other globulars. It's surprising that M15 is so bright, as it's nearly 40,000 light-years away. M15 is also the source of intense X-ray emissions, which indicates there may be something violent going on near its very compact core.

M2 is actually brighter than M15, at 6.0, but not quite so compressed as M15, so it isn't as impressive. M2 is 50,000 light-years away. If we could move it closer, it would vie with M13 and M22 for our attention. As it is, it's no slouch.

M71 is either a very compressed galactic cluster or a particularly puny globular cluster. It has features that suggest it could be placed in both categories. At magnitude nine, it's certainly not a showpiece object. However, it's sufficiently unusual to be worth a look, especially if you're in the vicinity looking at M27 just to its north.

Altair: New Meaning to the Phrase, "Whirling Dervish"

The eye of **Aquila**—the Eagle, is Altair a white star of magnitude 0.8, about 16 light-years away, about nine times brighter than our Sun. It's not of any particular interest to observe Altair, and its companion, at a distance of 165", isn't physically linked to Altair. They just happen to be almost in the same line of sight with each other. Why, then, is it remarkable?

Altair is a veritable whirling dervish. Whereas our Sun takes a leisurely 25 days to rotate once on its axis, Altair does that in 6.5 hours. It's rotating about 92 times faster than the Sun. The price Altair pays for this is that it must wear a flat-top. No, not the fifties hairdo that's coming back into style. Altair is only half-again as thick as it is wide, all the result of its absurd rotation rate.

About the ALL-Sky Map for September

September is another transition month. The northern constellations of summer remain high in the sky, but the "trademark" constellation—Scorpius—has already vanished, and the Teapot is low in the southwest. On the "other side of the sky," harbingers of the bright winter constellations are just clearing the northeastern horizon. But there's a lot more....

Vega Leads the Western Path

To find September's stars, begin by facing due west. You're looking for a bright-white star that's almost due west and about halfway between the horizon and the zenith. The key to finding this star—Vega—is that it's one of three bright stars that form

ALL-Sky Map for September

Map Symbols

●	−1
●	0
●	+1
•	+2
•	+3
·	+4
·	+5

a huge triangle in the sky west of the zenith. The others are Altair, off to the left of Vega, and Deneb, which you'll find above it, just below the zenith.

If your sky's dark enough to see the Milky Way, deciding which one of these bright stars is Vega will be relatively easy. Vega's the one that's NOT embedded in the Milky Way. In this view, it's between the Milky Way and the western horizon. If you can't see the Milky Way, just remember Altair is the dimmer star far to the left of Vega, and it's flanked on either side by stars of third and fourth magnitudes. Deneb, Vega, and Altair make up the Summer Triangle, a supergroup of stars. We'll get back to them later.

Once you're sure you've targeted Vega, you can make double sure by noting that there's a parallelogram of dimmer stars just to its left. With

Vega, these stars form the constellation **Lyra**—the Lyre, an ancient stringed instrument.

Just below and to the right of the Lyre, you'll notice a trapezoid-shaped group known as the Keystone, which the ancients knew as **Hercules**—the Strongman. The stars to the north and south of the Keystone are supposed to be Hercules' legs and arms. If you can find your version of Charles Atlas in these stars, add to the figure of the Keystone!

Now return to Vega and face a bit toward the southwest, putting Vega to the right of your field of vision. Just above Vega you'll return to Deneb, the topmost star in the Summer Triangle in this view. Starting with Deneb, trace the Northern Cross, which is the inner part of **Cygnus**—the Swan. The full figure of the Swan is found on this month's ZOOM map.

In this view, Altair is the bright star halfway between the zenith and the southwestern horizon. Altair is the "eye" of **Aquila**—the Eagle. In tracing the eagle, note the two stars that flank Altair. If they're not there, you've "latched onto" the wrong star!

If you have a dark sky and the Milky Way is visible, notice how it meanders through Cygnus and Aquila. You may notice a bright cloud at the bottom of the Eagle. That's the **Scutum Star Cloud,** actually a hole in the dark dust lane between the spiral arm "we live in" and the Sagittarius arm that's "one-arm-in" from us. The dust lane is visible as a split in the Milky Way that begins just above the Scutum Star Cloud and continues all the way through Cygnus to near Deneb. Some people call this split the **Altair Rift.** See if you can trace it.

If your sky is dark and your horizon is very low, be sure to look very low on the southwestern horizon to say farewell (for the summer) to the Teapot, aka **Sagittarius**—the Archer. If you haven't such a good horizon, or your sky isn't the darkest, go out on another night this month an hour earlier than the time for this map, about 09:20 P.M. That'll put the Teapot quite a bit higher.

To the left and well below Aquila, you should be able to trace the figure of **Capricornus**—the Sea-Goat. This figure reminds me of a sea-lion, but could also be called "Mick's Lips," if you happen to be a fan of rock stars as well as sky stars.

To the east of Capricornus, almost all the way around the horizon to due east, the sky is devoid of bright groups of stars. The constellations that occupy this area are only of interest to the people who invented them! They are, however, not totally devoid of interest, as you'll find when you read the commentary on the October ALL-Sky map.

What the Dippers Are Doing

If you're reading this in a place north of about 35° north latitude, you'll see the **Big Dipper** just to the left of due north, and very close to the horizon. If you live south of latitude 35° north, the Dipper is pretty much cut off by your northern horizon. Bright skies or a high horizon will make it nearly impossible to see the Dipper just now, so you'll have to find the "other" dipper using alternate means.

The "other" (official) dipper is the **Little Dipper.** To find it, you'll first need to look a bit to the right of due north, and about two-thirds of the way from the horizon to the zenith. There you'll find a group of stars in the shape of a squashed 3. Many people see this as a squashed M that's lying mostly on its side. In any case, it's the bottom of the figure that's "squashed" in this view: **Cassiopeia**—the Queen in Her Chair.

In the fall, Cassiopeia is the brightest constellation that's both close to the North Celestial Pole and well above the horizon. Just to her left, and due north, you can trace the outline of **Cepheus**—the King of Ethiopia. To check that you've found old Ceph, continue to face north, but crane your neck back so you can see the zenith. Just above and to the left of Cepheus you'll find Cygnus. Under dark skies, Cygnus is deeply embedded in the northern Milky Way, which extends into Cepheus and Cassiopeia and then to the northeastern horizon.

Just below Cepheus, you'll find the second-magnitude star Polaris, which guards the North Celestial Pole. A delicate arc of fourth- and fifth-magnitude stars extends to the left of Polaris, ending in the box which is the bowl of the Little Dipper. In finding the Little Dipper, remember that the two "end stars" of the dipper are considerably brighter than the stars between them and Polaris.

A Short Eastern Path

To see the fall constellations as they're rising, face due east. The key is to refind Cassiopeia, which you also used to find the constellations near the North Celestial Pole. In this view, Cassiopeia is a squashed W. It's a bit to the left of due east, and two-thirds of the way from horizon to zenith. Once you believe you've found it, verify this by tracing Cepheus, just above the W, and Cygnus (or the Northern Cross) beyond Cepheus. Remember, you'll have to crane your neck well past the zenith to see Cygnus.

Once you've refound Cassiopeia, look well to its right to trace the outline of the **Great Square of Pegasus.** It's really a rectangle, and it's relatively big, being 15° from north to south, and 18° from east to west. The way you're looking at it now, the Great Square is a bit taller than it is wide.

A graceful arc of stars extends from the lower left star in the Square: **Andromeda**—the Chained Lady. As far as the "Konstellation Kops" are concerned, the corner star of the Square is part of Andromeda. However, since almost everyone I know sees the Great Square, I've drawn it that way. Also, many people see the two as a "Great Dipper." You might, too.

Just to the right of the Great Square lies the lovely little group known unofficially as the **Circlet,** part of **Pisces**—the Fishes, one of those constellations of the Zodiac that I find impossible to trace. Most of Pisces lies below the Square and the Circlet.

Another invisible Zodiacal constellation is **Aquarius**—the Water Bearer, to the right or south of the Circlet. I've only drawn the Water Jar in Aquarius, as that's all I can see.

Below the handle of the "Great Dipper" lies **Aries**—the Ram, usually drawn as I have drawn it. What's really unusual is that I see it that way. (I'm compelled to point out, however, that a constellation whose figure consists of connecting two bright stars could be anything, including the Treasure of Sierra Madre, the Hunchback of Notre Dame, or Elvis Presley, complete with his guitar and gold records.)

At the end of the handle of the "Great Dipper" or Andromeda, you'll find the vaguely arrow-shaped figure of **Perseus**—the Hero. He's associated with Andromeda, Cassiopeia, and Cepheus in an episode of the cosmic soap-opera that's called Greek mythology. I'll tell all the juicy details in the commentary for the November ZOOM map.

Just below the figure of Perseus glitter the **Pleiades,** the perfectly wonderful galactic cluster of stars that herald the coming winter. Although I'm never overjoyed at the prospect of winter's arrival, I'm always heartened by my first view of the Pleiades each fall.

About the ZOOM Map for September

It may not seem so, but I have a strategy in picking what area of the sky to devote to which month's ZOOM map. It's to zoom in on an interesting area in the month when that area is best seen. If I followed that strategy to the letter, July would have three Zoom maps, and August and September would have none. That's because the region from Cygnus to Scorpius is really best viewed at the dates and times assigned to the July map. Because the Scorpius/Teapot region is far to the south, it has to be July's feature attraction every year, because it doesn't get very high above our horizon anyway.

The region of Scutum, Aquila, and Capricornus was just past the meridian in August, so it was still high in the sky in that month. That argued for it to be August's centerfold.

This month's target, the Cygnus/Lyra region of the northern Milky Way, is also a bit over the hill, but it's still quite high in the sky. Because it is, Cygnus/Lyra is easily observed with binoculars. You can orient your lawn chair facing west, tilt it so you're lying almost flat, and rest your binoculars on the bridge of your nose to steady them. Even though you'll be very comfortable in your lawn chair, the objects I'll be helping you find are guaranteed to keep you awake!

A Swan Afloat in a Sea of Milk

Pardon my waxing a bit poetic, but the figure of **Cygnus**—the Swan is one of the nicer constellation

ZOOM Map for September

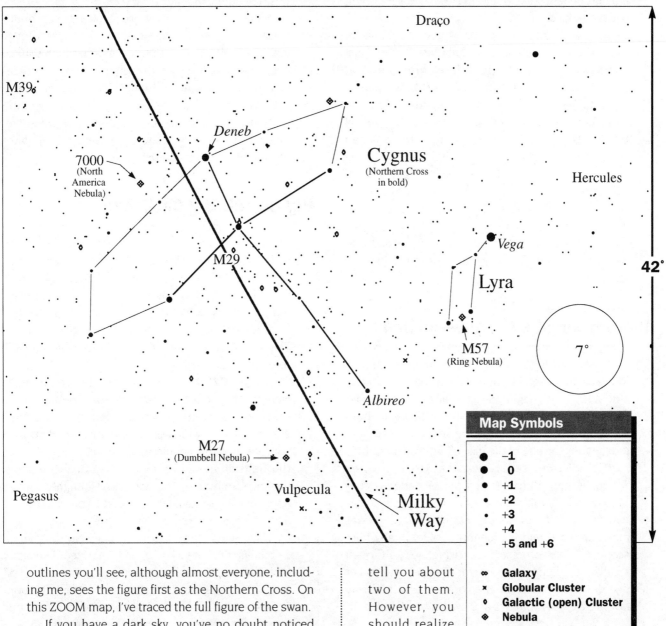

outlines you'll see, although almost everyone, including me, sees the figure first as the Northern Cross. On this ZOOM map, I've traced the full figure of the swan.

If you have a dark sky, you've no doubt noticed that the Milky Way runs along the swan's body, with a rift of darkness running to the southeastern side. This is the continuation of the Altair Rift that begins just north of the Scutum Star Cloud. It ends near Deneb at the swan's tail. According to the most accepted model of our galaxy, we're looking along the Orion Spiral Arm when we look toward Cygnus and that accounts for all the galactic clusters we see near the swan.

Those Galactic Clusters

The open clusters we see in Cygnus aren't the showpiece objects some others are. In fact, I'll only tell you about two of them. However, you should realize there are many more clusters than space lets me include. You might even think that the Milky Way running along the body of the swan constitutes one big galactic cluster. Indeed, it does contain our first one.

M29 may be found a bit south of the intersection of the Northern Cross within the Altair Rift. It's a nice cluster of ten to twelve stars, all of the eighth or ninth magnitude. This cluster would be much brighter than it is if it weren't for the galactic dust astronomers have found impeding our view. There's so much dust that for every photon that gets to our eyes, fifteen are scattered by the dust grains, which means that without the presence of the dust, these stars would be fifth or sixth magnitude, and that would make M29 a very bright cluster, indeed.

For those dozen or so stars to be that bright at their distance of about 7000 light-years, they must intrinsically be very bright. They are. The total brightness of M29 is possibly 50,000 times that of our Sun. We see it as a seventh-magnitude cluster.

M39 is found about 9° east and a little north of Deneb. Unlike M29, which hangs out in a bad neighborhood, M39 lies in a much richer field of the northern Milky Way. It's a much larger, looser collection of about thirty stars. It's only about 800 light-years distant and hasn't as much dust dimming our view of it, yet M39 is only two magnitudes brighter than M29. If you guessed that M39's stars aren't the powerhouses found in M29, you'd be right.

The Boy Scout's Delight—Albireo

I know of no double star that surpasses the one found at the nose of Cygnus—the Swan. It's Albireo, a double star of magnitude 2.9 that's so separated it can easily be split by binoculars if they're held steady. The two components are about 35" apart, approximately the size of the disc of the planet Jupiter.

It's not their separation that distinguishes Albireo A and Albireo B, it's their wonderful contrasting colors. Albireo A is a beautiful yellow star of magnitude 3.1. Its distant companion shines at magnitude 5.1, and is a beautiful blue. In any telescope, these two present a striking view, although small-to-medium-sized telescopes at very low power do better than their bigger brothers. In fact, my favorite view of Albireo is at 24× in a 4-inch reflector.

Lyra's Exquisite Ring

Let's leave Cygnus for a bit to describe one of the true showpiece objects in the northern sky, the Ring Nebula in Lyra. This object, known more formally as **M57,** is halfway between the southern leg of Lyra's parallelogram.

The Ring is a planetary nebula, those that are formed when a star rather peacefully sloughs off a large part of its outer layers when it becomes unstable late in its life.

Since it's a ninth-magnitude object, you might think the Ring is dim. However, its size is fairly small, only 60" by 80", so it has a rather high surface brightness. A 3-inch will only show it as a disc-shaped object, while a 4-inch will sometimes show the ring shape. A 6-inch shows the Ring Nebula quite nicely. Of course, in my 20-inch it's awesome. Since it's so small, you'll want to use medium power on the Ring Nebula. Enjoy!

Big Object, Big Challenge

The first time you look for it, you may not see the last object I'll describe. Depending on your dedication, skill, luck, the humidity, and the darkness of your skies, you may not find it after dozens of tries. With the exception of the Horsehead Nebula, it may be the most difficult object to find that I'll tell you about. However, I can't resist. It's too beautiful NOT to try to find.

It's **NGC7000,** known as the North America Nebula. This object is a huge emission nebula in the shape of the North American continent. It lies 3° east of Deneb, which is sufficiently bright to ruin your view of the nebula if you don't keep it out of the field of view. (Since Deneb is also the same distance from us as the nebula, 1600 light-years, it appears it's also sufficiently bright to provide the UV light that excites the nebula.)

There are several reasons why the North America Nebula is difficult to find. One is that people usually look for a much smaller object than is there. It's 1.5° across, or three times the diameter of the full Moon. That's about 20 percent of the diameter of your binoculars' field. So you really must use a very low power in a small telescope to see NGC7000. Second is the surface brightness of NGC7000. Although its total magnitude is about four, that amount of light is spread over a large area. To maximize contrast, you must use your absolutely lowest power. Also, if the Moon's up, or it's very humid or hazy, or you're not out in the boonies, looking for the North America Nebula is a waste of time.

About the ALL-Sky Map for October

Octtober is another of those months that feature a transition from the gaudy summer sky to the almost-as-gaudy winter version.

Watching the Summer Triangle Sink in the West

Getting a handle on the stars of autumn begins with the bright **Summer Triangle.** To find it, face just a bit north of due west. About 25° above the western horizon, you'll find Altair, a bright white star with a third-magnitude star to its right, and a slightly dimmer fourth-magnitude star also to its right. Well to Altair's right and a bit higher lies Vega, brightest star in October's sky. Above Vega and to its left you'll find Deneb, which completes the Summer Triangle.

Now trace the Summer Triangle back to Altair. This bright star is the eye of **Aquila**—the Eagle. You can trace his figure as he appears to be standing near the horizon spreading his wings.

If you have a dark sky, you'll notice that the Milky Way runs right along the Eagle's body right up into **Cygnus**—the Swan. To make it less confusing to find Cygnus, I've drawn it as the more familiar Northern Cross. Like the Eagle, it's standing almost straight up from the horizon. (The dark area that you can see running from the right of Altair up into Cygnus is called the Altair Rift. Just to its right, the Cygnus Star Cloud runs right up the body of the Northern Cross. This bright part of the Milky Way is an area within our own spiral arm of the galaxy.

Vega is the brightest star in the constellation **Lyra**—the Lyre. Besides Vega, the Lyre is made up of the dim parallelogram of stars that extend a bit below and to the left of Vega.

To reinforce what you've just learned around the Summer Triangle, trace the figures of Lyra, Cygnus, and Aquila as they prepare to set. For the sake of completeness, you can look to the left of the Eagle for the dim figure of **Capricornus**—the Sea-Goat.

What the Dippers Are Doing

Mostly, they're hiding. That's because the **Big Dipper** is in that part of its journey around the sky that's either at or near the northern horizon. For the latitude for which this map was drawn, 35° north, the Dipper is just grazing the northern horizon. Only if you have a very low northern horizon and live north of latitude 35° will you likely see the Big Dipper. If you live south of 35°, all or part of the Big Dipper will be below the northern horizon at the time of October's map.

Without the Big Dipper to point to it, the **Little Dipper** is fairly hard to find. The easiest way is to find all the other major constellations around it. Start by facing north. About two-thirds of the way up from the horizon to the zenith you'll find a group of stars in the shape of a squashed M. This is **Cassiopeia**—the Queen in Her Chair. If your sky is dark, you'll notice the Milky Way runs right through Cassiopeia. Just below and to the right of Cassiopeia lies her husband, **Cepheus**—the King of Ethiopia.

Just below the lower tip of Cepheus, you'll find a third-magnitude star pretty much all by its lonesome. That's Polaris, the North Star, which is less than a degree from the North Celestial Pole. Polaris is at the end of an arc of stars that drops down and to its left to form the handle of the Little Dipper. The small box of stars at the end of the arc forms the bowl. This is not a constellation that jumps out at you. It helps to remember that the bottom two stars in the bowl are considerably brighter than the others between them and Polaris.

Now work your way back up through Cepheus to Cassiopeia. You'll use the Queen to lead you to the constellations in the northeastern part of October's sky.

N

Big Dipper

Little Dipper

Hercules

Cepheus

Gemini

Auriga

Cassiopeia

Lyra

Perseus

Cygnus

Andromeda

Pleiades

E

Taurus

Hyades

Aries

Great Square

Pegasus

W

Orion

Pisces

Aquila

Circlet

Mira

Aquarius

Cetus

Water Jar

Capricornus

Fomalhaut

S

Map Symbols

●	**−1**
●	**0**
●	**+1**
•	**+2**
·	**+3**
·	**+4**
·	**+5**

The (Dim) Stars of Autumn to the (Bright) Stars of Winter

While you're still looking at Cassiopeia, slowly shift your body so you're looking due east. Cassiopeia's squashed-**M** shape now looks either like a squashed **3** or a squashed **W** on its side, depending on your whim at the moment.

Immediately to the right of Cassiopeia, a bit past the zenith, you'll find the huge square-shaped group of stars known, amazingly enough, as the **Great Square of Pegasus.** To best see the Great Square, just lie back in a reclining lawn chair so you're looking straight up. To the ancient inventors of the constellation figures, Pegasus is a Winged Horse.

As you continue to face east, you'll notice a long shallow arc of stars extending down and to the left of the Great Square. That's **Andromeda—** the Chained Lady. The star to which the arc is attached is within the constellation boundaries of Andromeda, while most of the (alleged) Winged Horse lies on the opposite side of the Square from Andromeda.

Many people see the Pegasus/Andromeda figure the way I've drawn it, as an enormous dipper. If it helps you to see it that way, then see it that way. If you see a winged horse and a chained lady in those stars, you get a lifetime A+ for imagination.

Speaking of imagination, you'll need some to recognize the unrecognizable. Just below the Great Square lie **Aries—**the Ram, **Pisces—**the Fishes, and **Aquarius—**the Water Bearer. So you'll know where they are, I've drawn the "figures" of Aries and the Water Jar in Aquarius. One feature I do see, and you might, too, is the **Circlet.** This little almost-circle of stars is part of Pisces, but is wedged between Pegasus and Aquarius.

The region of Pegasus/Andromeda is featured in this month's ZOOM map. In the commentary, I'll introduce you to one of the most breathtaking objects in the heavens, the Great Andromeda galaxy. As a bonus, you'll also get to find out how

the lady Andromeda found herself in chains and then in the sky!

To finish up October's skies, look just off the handle of the great dipper formed by Pegasus/Andromeda. There you'll find **Perseus—**the Hero. His vaguely arrow-shaped figure also lies in the heart of the Milky Way, but now we're looking through another hole in the dust lanes of our galaxy into the spiral arm that's one farther out from ours. Because our hero's stars are mostly in that arm, it's called the Perseus Arm. (Perseus and Cassiopeia are featured in the November ZOOM map.)

Below and to the left of Perseus you'll find the figure of **Auriga—**the Charioteer. He's captained by the beautiful reddish-yellow star Capella, traditionally one of the heralds of the arrival of the stars of winter.

Just below and to the right of Perseus' feet lie the **Pleiades,** clearly the most beautiful of the heralds of the approach of winter skies. Directly below them lie the **Hyades,** another large, bright galactic cluster. Along with the bottom star in the pentagon of Auriga, the Pleiades and the Hyades form **Taurus—**the Bull. No one actually sees this figure. Rather, they see the Pleiades and the Hyades as separate groups. That's the way I've drawn them. The region around the Pleiades and Hyades is featured in December's ZOOM map.

About the ZOOM Map for October

October's ZOOM map is devoted to a pair of bright, close galaxies. However, there's also a bright, colorful double star, and the story of Perseus and Andromeda. (Despite what it may seem, this last isn't the subject of a made-for-TV movie designed to be watched by football-widows on Monday nights.)

The Exquisite Andromeda Galaxy

Our planet, Earth, is not located exactly in the plane of the Milky Way. Rather, we're a bit "below" the exact center of the galactic disc. This allows us to look a bit under the dust lanes that lie between

ZOOM Map for October

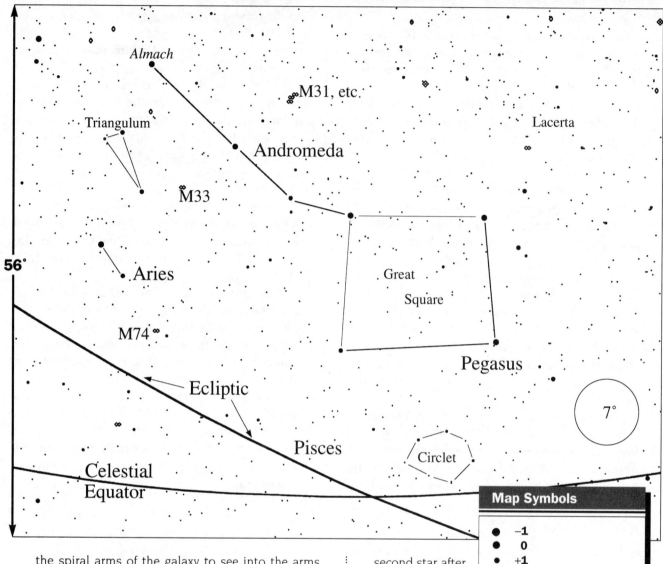

the spiral arms of the galaxy to see into the arms inward from us. It also allows us to easily look out of our galaxy to see the **Great Andromeda galaxy,** the flagship of our Local Group. It's M31 on Messier's list of "fuzzies."

We're indeed fortunate to be where we are, because we'd be much the poorer for missing the view of the Andromeda galaxy. It's that good. It's so close that it's bright enough—magnitude 3.5—to be seen without optical aid, if you're searching for it on a dark, moonless night.

Here's how: Start at the northeast "corner" of the Great Square of Pegasus. You'll remember that the constellation Andromeda forms the "handle" of a great dipper, with the Great Square as its bowl. Now proceed along the handle to the second bright star. It's the last one before the end of the handle. Now make a "right turn" toward Cassiopeia. Just past the

second star after the "turn," you'll notice a "fuzzy star." That's the Andromeda galaxy.

This first "naked eye" view may not impress you much. However, think about what you're seeing with just your own eyes—a galaxy. It's not an object within our own galaxy. Nor is it one of our (close) satellite galaxies that our friends in the southern hemisphere are privileged to see. No, it's *another* galaxy.

By the distance standards of our own galaxy, the Andromeda galaxy is quite far away, about 2.3 million light-years. So, we're seeing it not as it is

Map Symbols

Symbol	Magnitude
●	−1
●	0
●	+1
●	+2
•	+3
•	+4
·	+5 and +6

∞ **Galaxy**
× **Globular Cluster**
○ **Galactic (open) Cluster**
⬦ **Nebula**

now, but as it was 2.3 million years ago. It's a lot bigger than our galaxy, being about 160,000 light-years across. It also is quite massive. It has the mass of about 300 billion stars like our Sun. However, since there are about 400 billion stars in this galaxy, the average star isn't as massive as the Sun.

The Andromeda galaxy only puts out as much light as 13 billion of our Suns would emit, which might seem to be a bit weak. It's easily explained when you consider that the many stars in the galaxy are not particularly luminous.

When you look at the Andromeda galaxy in a small telescope, it becomes obvious that the majority of the light it emits is coming from the core. It's extremely bright, looking like an enormous globular cluster that's not been resolved. The remainder of the galaxy, the system of spiral arms, is considerably dimmer.

The first time you look at it, the core may seem to be just a bright white in color. However, you should notice what color it's *not*. It's not the greenish-white of a bright emission nebula. If you're looking at it in a small telescope, it's probably going to just look white. However, if you can get a look at it in a larger telescope, it may take on a slightly yellowish cast.

The reason the core looks slightly yellowish is that most of the stars in the core are old red stars that formed soon after the Andromeda galaxy coalesced early in its life. Since these stars are in a region where there's little gas and dust from which new stars can form, there are no new ones putting out bluish light to "balance" the spectrum of the core. At low light levels, our eyes can't see much of the red end of the spectrum, so we see the core as a bit yellowish.

The stars that have formed out in the spiral arms of the Andromeda Galaxy are younger bluer stars. Since there's still a lot of gas and dust in the spiral arms, new stars will be forming there for many billions of years to come.

The core is bright but featureless, however, the spiral arms that make up the bulk of the Andromeda galaxy are dim and full of features. A 6-inch telescope will reveal two dark dust lanes on the west side of the galaxy. However, this only happens under extremely dark skies when the galaxy is high in the sky. If you have a dark, humid night, you may not see the lanes at all in a 6-inch, or even an 8-inch, as the humidity will spoil the contrast. A 10-inch will show them on most really dark nights. Of course, much larger telescopes show much more of the dust lanes, but, since their field is limited, you have to "walk" the telescope around the Andromeda galaxy to see it all.

The most important factor in seeing the dust lanes of the Andromeda galaxy is to use the lowest power you can. That will ensure that you'll have the highest possible contrast. If you're a beginner, you might not notice the dust lanes the first few times you look for them. Keep plugging, though. As your eye-brain combination gets used to looking at very low contrast objects, you will.

The dust lanes are not the only feature in the spiral arms you can see. There is one star cloud that's so bright that it has its own designation in the *New General Catalogue* of nonstellar objects— **NGC206.** You'll find it near the south end of the Andromeda galaxy.

Although they can't be observed in typical amateur telescopes, you'll probably find it interesting that astronomers have used the great observatory telescopes to observe that the Andromeda galaxy contains all the familiar objects we observe in the Milky Way. Not to be outdone by the Milky Way, which has two satellite galaxies, the Andromeda galaxy has four such satellites. All four are dwarf elliptical galaxies and have almost no young stars in them. Only two of these companions are close to M31 in the sky, **M32** and **M110.** M32 is due south of the core of M31 and shines at magnitude 9.5. It's about 3' across, so it mostly looks like a fuzzy star at low power. M110 is bigger and dimmer, so you might find it difficult to find in a small telescope. Its magnitude is 10.8, and it's 3' by 8'.

Don't forget to look at the Andromeda galaxy in binoculars. When you do, you'll see the bright core surrounded by the lens-shaped spiral arms. Although the "official" size of M31 is 1 by 2.7°, many observers have seen it extend to as much as 4° when viewed in binoculars. I'll never forget a view of the Great Andromeda galaxy I had in a 4-inch low-power telescope I built when I was a kid. It was literally breathtaking. I hope your first view of Andromeda will be as memorable.

Another Beautiful Member of the Local Group

Except for the fact that it contains the galaxy **M33,** the constellation **Triangulum**—the Triangle would be somewhat of a washout. However, the subtle M33 saves the day. This loose, almost-face-on spiral may be found on "the other side" of the handle of the great dipper from the Andromeda galaxy.

Where M31 shouts its presence with its large size and brilliant core, M33 prefers to whisper. Its loose structure and small size mean it's much more difficult to observe than its flashy neighbor. So, on many nights, you'll find it difficult or impossible to observe. On those nights when M33 is easily visible, you'll find other objects show you much more than you usually see in them. So, M33 is something of a bellwether for the transparency of your sky. As always, low power is the key to observing M33.

M33 is only about 40 percent the size of M31, or about 65,000 light-years across. At its distance of about 2.7 million light-years, it ends up being about a degree across "on the sky." Its magnitude 6.5 brightness is spread out over quite an area, explaining how dim it often appears.

Back to Andromeda for a Bit More

The "end star" in the great dipper formed by Andromeda and Pegasus is called Almach. It's a beautiful, wide-spaced double star, second in beauty only to Albireo, the wonderful blue-and-gold double star in Cygnus. Almach A is a 2.2-magnitude golden yellow beauty. Its magnitude 5.1 companion lies 10" away and is bluish-green. Since the separation is so great, you'll have little difficulty separating Almach into its components. To see the contrasting colors at their most vivid, keep the power only high enough to separate cleanly the two stars.

A Cosmic Soap Opera

Andromeda was the only daughter of Cepheus and Cassiopeia, the King and Queen of Ethiopia. Mama Cass had a thing about Andromeda's beauty, and spent so much time extolling her daughter's charms that she managed to anger the sea nymphs. (Other sources claim the problem resulted from a dispute over fishing rights.)

Whatever the provocation, it appears the Sea Nymphs persuaded King Neptune to punish Cassiopeia by allowing Andromeda to be chained to the rocks next to the sea. The object of this exercise was to provide the local sea monster with a tender morsel of Princess on which to dine.

Perseus had other ideas. He appeared at the rocks with the head of Medusa. The monster took one look at Medusa, and was instantly turned to fiberglass, cut up, shipped to a small town in the Ozarks, reassembled, and made the subject of a roadside "dinosaur museum." Perseus became a hero, married Andromeda, and as his reward for saving her, was placed in the sky with her AND the in-laws. Pretty good cosmic-soap, wouldn't you say?

About the ALL-Sky Map for November

November's ALL-Sky map looks busier than it is. That's because I've added two constellation figures to this map that will not appear on any of the others. I've drawn Cetus—the Whale to help you find the variable star Mira and the bright galaxy NGC253, which are bonus objects this month.

I also added Eridanus—the River. I drew its outline solely to illustrate the Greek Konstellation Kops' First Law of Sky Figures: If you have a jumble of stars, and you're really stuck for a constellation fig-ure to make out of them, you can always make it into a river. (There's a corollary: If you don't have quite enough stars for a respectable river, try a serpent.)

Eridanus and Cetus, among other nondescript constellations, occupy the southern third of November's sky. For beginning observers, there's just not a lot there, so I'll not take you through them. On that note, let's get started on the constellation figures reasonably observant humans such as you and I can see....

ALL-Sky Map for November

N

Big
Dipper

Little
Dipper

Lyra

Cepheus

Cancer

Cassiopeia

Cygnus

Auriga   Perseus

Gemini

Andromeda

E   Canis
    Minor

Aquila   W

Great
Square   Pegasus

Pleiades

Taurus   Aries

Orion   Hyades

Pisces

Aquarius

Circlet

Mira

Water Jar

Canis
Major

Capricornus

Eridanus
the (what else?) River

Cetus (the Whale) 253

Fomalhaut

S

Map Symbols

●	−1
●	0
●	+1
•	+2
•	+3
·	+4
·	+5

What the Dippers Are Doing

Just as in October, the Dippers are mostly hiding. We'll find them, but not without some help.

Although it's now late fall, you might try continuing to use a reclining lawn chair for your constellation study. You can keep warm by snuggling up in a sleeping bag.

Face you and your recliner due north and lean back. About two-thirds of the way from the horizon to the zenith you'll find a group of stars in the shape of a squashed M: **Cassiopeia**—the Queen in Her Chair. If you're observing under dark skies, you'll be able to see the Milky Way running through Cassiopeia as it works its way from just north of the western horizon to just south of the eastern horizon.

Just below and to the left of Cassiopeia, you'll find the large, somewhat dim figure of **Cepheus**— the King of Ethiopia. Just to the right and a bit below the King's point lies the second-magnitude star Polaris, the North Star. Polaris is less than a degree from the true North Celestial Pole, so it hardly moves in its daily journey around the sky.

If you look carefully, you can see a faint arc of stars extending below and to the left of Polaris. At the end of the arc is a small box of stars. The lower two stars in the box are brighter than the others between them and Polaris. You've found the **Little Dipper,** which is a dim dipper, but an even poorer little bear, which is its formal designation.

The **Big Dipper** really hugs the northern horizon in the fall. If you live at or north of 35° north latitude, you can just see most of the Dipper stretched out along the horizon to the right of due north. Of course, if you live farther south, the northern horizon cuts off some or all of the Dipper this time of year. You'll also not see it unless you're observing from a dark-sky site with a nice, low northern horizon.

Retrace your steps back to Cassiopeia. We'll use her figure as the point of departure in finding the rest of November's stars.

A Path to the Cross and Then Back Across (the Sky)

Once you've traced your way back to Cassiopeia's squashed **M**, look way over to the left past Cepheus. There's a bright white star about a third of the way from the horizon to the zenith. Extending below it toward the horizon is the figure of the Northern Cross, formally called **Cygnus**—the Swan. (In this part of the map, the constellation figures are distorted because the spherical sky can't be perfectly projected onto the flat page. The legs of the Northern Cross actually are not at as acute an angle as the map shows them.)

The stars of Cygnus are the last of the stars of summer. As you bid them farewell, retrace Cygnus and go past Cepheus back again to Cassiopeia.

Now swing around to face south. Lean waaaaaay back to once again pick up Cassiopeia. Now, of course, she's a squashed **W**. South of Cassiopeia,

very near the zenith, you'll notice a long arc of stars trailing off toward the west. This arc is about 20° long or twice the apparent long-length of your closed fist when you hold it at arm's length.

At the end of the arc, you'll notice a large square of stars. Most people see the arc and the square as a "Great Dipper." The arc of stars that makes up the "handle" of the great dipper is **Andromeda**—the Chained Lady, while the big square of stars that makes up the bowl is the **Great Square of Pegasus.** The star located where the handle joins the bowl is actually part of Andromeda, but is usually associated with the Great Square. That's probably a good idea, since the Great Square is easy to see, while its formal figure, the Winged Horse, is thoroughly unrecognizable.

Below Andromeda lies **Aries**—the (thoroughly unrecognizable) Ram and to its right is **Pisces**—the (thoroughly unrecognizable) Fishes. Just below the Great Square of Pegasus is a nice little asterism known as the **Circlet.** It has no formal recognition, but people can recognize it, so I drew it in. Past the Circlet lies the (thoroughly unrecognizable) **Aquarius**—the Water Bearer. I was born under its astrological sign, but I've never been able to conjure up enough imagination to see a Water Bearer lurking in those stars. You'll probably be able to find the Water Jar, though.

To get off our thoroughly unrecognizable kick, find the Circlet again, then the Great Square and its handle, Andromeda. As you're still looking south, Cassiopeia will be above Andromeda. Now swing around to face due east. Look at Cassiopeia as you do, so you won't lose your reference.

Below Cassiopeia and Andromeda you'll see the form of **Perseus**—the Hero. (I told the story of Perseus and Andromeda in the commentary for the October ZOOM map. Perseus and Cassiopeia are the subject of this month's ZOOM map.)

Just below Perseus lies **Auriga**—the Charioteer. You can recognize him by the distinctive pentagon shape (of the charioteer?). The bright star Capella, which I see as slightly reddish-yellow, anchors this figure.

Two magnificent galactic clusters lie just to the right of Perseus and Auriga. The higher one is the magnificent **Pleiades,** a mini-dipper-shaped group that, to me, is among the most delightful sights in the sky. Below the Pleiades you'll find the **Hyades,** a much bigger cluster in the shape of an arrow. The bright-reddish star Aldebaran anchors the Hyades, although it's not actually a member of that cluster.

The Pleiades, Hyades, and the rightmost star in Auriga form **Taurus**—the Bull, one of the constellations of the Zodiac. I've spared you the frustration of trying to see a bull here by refusing to draw him.

Our last two constellations are figures of humans lying on their sides in the sky. Just below Auriga you'll find the two bright stars Castor and Pollux that mark the heads of **Gemini**—the Twins. Castor is on top, Pollux is below.

You'll also find **Orion**—the Hunter lying in repose above the horizon just south of due east. His distinctive belt of three bright stars and the bright stars that mark his shoulders and knees make him an easy group to find.

Bonus: Mira (the Wonderful) and (the Wonderful) NGC253

November brings us two classic variable stars. Algol is an eclipsing variable star that is the "star" of this month's ZOOM map. Mira is a long-period pulsating variable that's out in the middle of "sky nowhere," so it doesn't fit onto a ZOOM map.

Mira is too wonderful to be left out of this book, however, so I made it a bonus object. Using the word "wonderful" is appropriate, as that's what the name means in Arabic. The ancients gave it that name because to them it would appear and then disappear over and over again.

When we humans began looking at the sky through telescopes, Mira was the first star to be recognized as varying in brightness. Its magnitude varies between about three and nine over a period of about 331 days. However, the period isn't perfectly regular and can vary by as much as twenty days. Also, its maximum magnitude can be as bright as two.

Astrophysicists believe Mira is a red giant that's slightly unstable and pulsating in size. When it's at its biggest, it's also at its coolest, so much more of its energy output is in the far red and infrared part of the spectrum to which the human eye is not very sensitive. In that case, it appears very dim to our eyes. When Mira has shrunk to its minimum size, it's at its hottest, and more of its output is in the visible spectrum, so it appears brighter.

Of course, to an observer without optical aid, any object that is below magnitude 6 or 7 is invisible. Since Mira spends about half its period below the seventh magnitude, it "disappears" for half the year. If you have binoculars or a small telescope, Mira is an easy variable to follow. (To estimate Mira's brightness, see the commentary on this month's ZOOM map.)

Mira is part of a big dim constellation called **Cetus**—the Whale. On this month's map, I've drawn a fairly standard representation of Cetus.

Below and to the right of Mira you'll find the dim shape of the whale's jaw. Slightly south and west of his jaw, you'll find the spiral galaxy **NGC253**, a large object almost half a degree in length, shining at magnitude 7. If you live to the north, you might want to search for NGC253 about an hour earlier than the nominal time for this map, because it's at its highest at that time.

About the ZOOM Map for November

Your ZOOM view for November takes in the Perseus arm of our Milky Way. This spiral arm is "one out" from the arm in which we find ourselves. Part of what we see in Perseus and Cassiopeia is within our own spiral arm, and part is the Perseus arm we see through holes in the dust lane between two arms.

If the above description sounds easy, remember it took astronomers the better part of the twentieth century to figure all this out, and they don't all agree on the details, even today.

The classical deep-sky objects found in the Milky Way are the stars of this ZOOM map. However, a classical variable star leads off the list....

ZOOM Map for November

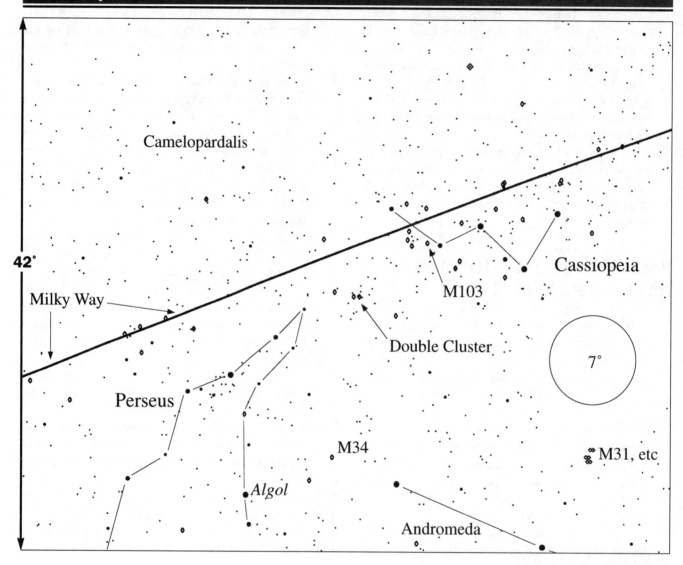

Algol, the "Demon Star"

The bright star at the foot of **Perseus**—the Hero varies in brightness between magnitude 2.1 and 3.4, a range of 3.3 times. It does this very regularly, with a period of 68.82 hours, or about 2.867 days. The "valley" of Algol's brightness lasts for about ten hours. Because they noticed its variability, the ancients apparently equated it with the Forces of Darkness, and called it Algol, which, loosely translated from the Arabic, means "Demon Star."

Early astronomers believed the variability in Algol's brightness was caused by a dark body partially eclipsing it as it rotated around it. This was the correct guess, except the "dark body" is really a much dimmer star.

Algol A is a white star of about 2.6 million miles in diameter. Its fainter companion is a reddish-yellow star that's actually bigger than Algol A. Algol B is 3 million miles in diameter, but it's so much cooler it's not nearly so bright as Algol A. So, when it passes nearly between us and Algol A, the companion causes the total magnitude of the system to drop from 2.1 to 3.4.

Map Symbols

●	−1
●	0
●	+1
●	+2
·	+3
·	+4
·	+5 and +6
∞	Galaxy
×	Globular Cluster
◇	Galactic (open) Cluster
◈	Nebula

Algol A and B are so close to one another we can't resolve them with Earthbound telescopes. Slight, regular changes in the period-of-variability of Algol have led astronomers to infer that Algol has a second companion with a period of about 32 years that orbits the close-eclipsing pair.

Observing Algol is easy. You can estimate its brightness even without optical aid, although binoculars or a small telescope will allow you to do a better job. You just compare its brightness to other nearby stars. If, for instance, you see that Algol's brightness is exactly between two stars whose magnitudes are 2.9 and 3.2, your estimate would be the average of the two, or 3.05. If Algol seemed to be closer in brightness to the brighter of the two comparison stars, your estimate would be 3.0.

Since this book is designed to get you started, I can't tell you all you might want to know about observing variable stars. However, the American Association of Variable Star Observers will be very happy to provide as much info as you can stand. They'd really like to hear from you. Their address is found in Section 8.2.3 of the text.

The Double Cluster and Other Wonders of November's Deep Sky

About halfway between Perseus's head and the "squashed" portion of Cassiopeia's W, you may notice a bright "fuzzy." (If your sky isn't dark, you may not see much of this without the use of binoculars or your finder.) Viewing this area through binoculars or a small telescope reveals it's a double cluster of stars or a pair of galactic clusters. Each of the objects in this pair is enumerated in the *New General Catalogue,* as **NGC869** and **NGC884,** but they're known as the "Double Cluster," so that's what I'll call them.

The Double Cluster defines the heart of the Perseus arm of our galaxy. It appears from studies of the spectra and evolution of the stars in each cluster that they are close to one another within the Perseus arm. Their estimated distance is 7000–8000 light-years, with NGC884 being the farther of the two. NGC884 may also be the older, although they're both thought to be very young, probably only a few million years old.

NGC869 has about four hundred stars associated with it. Three hundred stars seem to be members of NGC884. However, since some of these stars are as dim as 17th magnitude, you won't see all of them in

binoculars or a small telescope. Since they're separated by about 0.6°, you'll want to view them with a power that produces a field of no less than 1.0°. A larger field shows the Double Cluster against a beautiful field of Milky Way stars.

For instance, the view of 869 and 884 in my 4-inch Newtonian at 17× is truly inspiring. The contrast is so low across the 3.8° field that the stars of the Double Cluster resemble hundreds of jewels. Most of the stars in the clusters are bluish-whites of recent origin. However, a few red giants provide a nice contrast. It's a wonder this group isn't called "the Jewel Box," as a beautiful galactic cluster visible only to southern observers is named.

When I increase to 41×, the contrast goes down, but I can still observe both clusters in a 1.6° field that allows a fine view. You don't have to prefer either view. Each has its merits.

A Nice Group of "Other" Objects

I put "other" in quotes because I want to emphasize that these objects are only secondary because Algol and the Double Cluster are so interesting. Although one is challenging, they all stand on their merits.

The region of Perseus and Cassiopeia is laced with galactic clusters, as you might expect from knowing that we're looking into a spiral arm of the Milky Way. I'll tell you about four of them. If you spend more than a few minutes sweeping this region, I'm betting you'll recognize more.

In Perseus, **M34** is a bright galactic cluster 5° west and a couple of degrees north of Algol. This cluster of about eighty stars shines at magnitude 5.5, which means you should be able to detect it without optical aid on a clear, dark night. The only thing that keeps it from being easier to detect with the naked eye is that it's big—20' in diameter—or about two-thirds the diameter of the full Moon. Just as you did with the Double Cluster, you'll want to use low power on this object.

M34 is not part of the Perseus Arm, since its distance is about 1600 light-years. Also, it's much older than the Double Cluster, being about one hundred million years old.

In Cassiopeia, **M103** lies just east of the bottom of the "squashed" section of the W. It's a compact cluster of 25 stars, the brightest of which are between magnitude 7 and 9. One eighth-magnitude red star adds spice to this 7.4-magnitude cluster. It appears that M103 is part of the Perseus arm, as its distance is about 8000 light-years.

Although Messier failed to list it in his catalog, **NGC457** is actually a full magnitude brighter than M103. This group of sixty stars is even more distant than M103, yet it's brighter. That's because a few of its suspected members are extremely luminous stars. One of its members is so luminous its magnitude is just sufficient to be plotted by the map software on the ZOOM map by itself.

M52 is a seventh-magnitude galactic cluster that's made up of about a hundred stars. You won't see that many in a small telescope. For instance, a 4-inch at low power will show possibly two dozen stars within a diameter of 10', or a third the diameter of the full Moon.

There's one quite challenging object to view back in Perseus. It's the big, dim planetary nebula **M76.** This object is the dimmest of those that appear in Messier's Catalog, at magnitude 12.2. Its great size, 40" by 80", means its surface brightness is low. If you have a new 6-inch telescope, you might try your hand at finding M76. With an 8-inch it should be a bit easier. A 10-inch will allow you to see M76 with little difficulty.

All these estimates assume you have a clear, dark night. Also, since M76 is big for a planetary, keep the power low until you recognize M76 as a dim "fuzzy." Once you think you've found it, bump the power up to make the background sky darken and allow you to see the shape of M76, which is that of a small "dumbbell."

About the ALL-Sky Map for December

For the astronomer, December is the first true month of winter. Only one of summer's bright constellations is still around and it's sinking fast. The dimmer constellations of autumn congregate near the western horizon. Since it's winter, it's appropriate that the constellations of winter dominate the eastern half of the December sky.

In the northern hemisphere, December is the month when the Sun reaches its farthest point in the southern sky. The days are at their shortest, the Sun stands low in the sky at noon, and we wish for the long, warm days of summer. And yet, many different cultures celebrate some sort of major religious event in December. These events often share a theme of light and hope, despite the short, dark days. Maybe it's appropriate that the stars of winter are so bright and cheerful.

An Unusual Path to December's Stars

If you want the easy way to learn December's stars, you can use the path I describe for finding many of January's constellations. I chose to use a

different path for December because I believe you'll learn the whole sky better if you crisscross it in many different, overlapping paths.

This strategy mimics how you learn your way around a new town. You move in and learn the way to work, the grocery store, the bank, and a local church or two. During your travels around town, you get lost. As you're driving along an unfamiliar street, you accidentally notice a familiar landmark. The street on which you were lost is now familiar territory. So I hope it will be with the PATHs.

What the Dippers Are Doing

You begin December's session under the stars by facing due north and looking at the horizon about 30° east of north. There, standing on its handle is the **Big Dipper.** If you live north of 35° north latitude, the Dipper will be fully above the horizon. If you live south of that parallel, a bit of its handle will be obscured by the horizon.

Once you've found the Big Dipper, it's an easy task to use it to find Polaris. Just use the top two

ALL-Sky Map for December

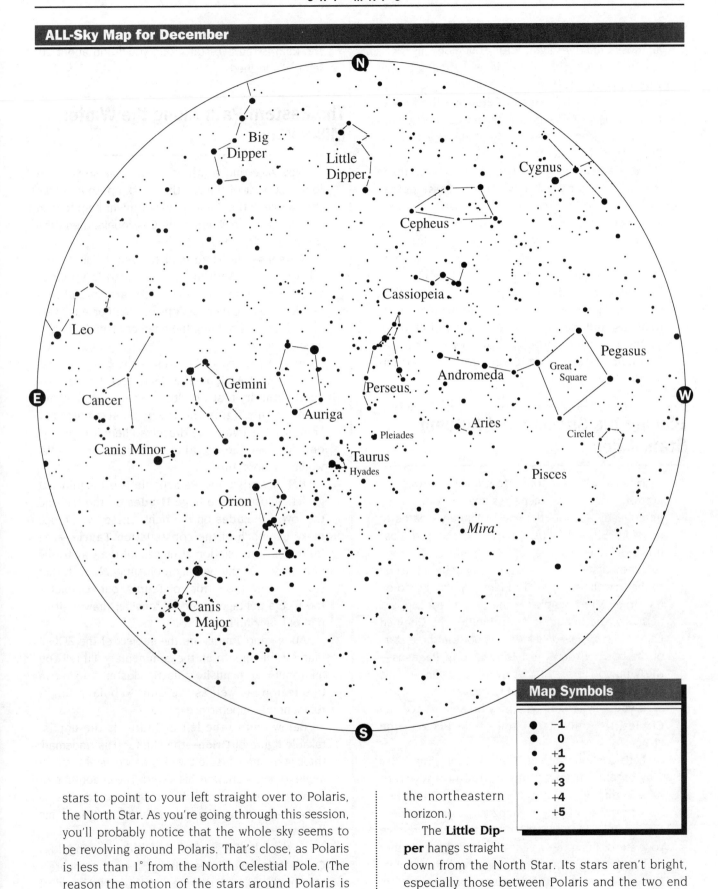

Map Symbols

●	−1
●	0
●	+1
•	+2
•	+3
·	+4
·	+5

stars to point to your left straight over to Polaris, the North Star. As you're going through this session, you'll probably notice that the whole sky seems to be revolving around Polaris. That's close, as Polaris is less than 1° from the North Celestial Pole. (The reason the motion of the stars around Polaris is especially noticeable in December is that it's easy to see the Big Dipper as it climbs away from the northeastern horizon.)

The **Little Dipper** hangs straight down from the North Star. Its stars aren't bright, especially those between Polaris and the two end stars in its bowl, but it has no competition in that part of the sky (except for the occasional source of light pollution, like the guy north of you who's got all those Christmas lights strung out along his roof).

On the other side of Polaris from the Big Dipper we run into celestial royalty. Just to the left of Polaris we find the figure of **Cepheus**—the King of Ethiopia. His silo shape points almost directly back at Polaris. Remember that Cepheus is just a bit bigger than the Little Dipper.

Above Cepheus you'll notice a bright group in the shape of a squashed M. This is **Cassiopeia**—the Queen in Her Chair. Old Cass wasn't a modest sort, as you'll gather from reading about her in the commentary on the November ZOOM map. She's relatively bright and with the Big Dipper anchors the region around the North Celestial Pole. If you look at each ALL-Sky map, you'll notice that Cassiopeia never goes below the horizon. She, Cepheus, and the Dippers are certainly blotted out by the Sun every morning, but they never suffer the indignity of having to set. They're called circumpolar stars.

Bidding the Summer and Autumn Stars Adieu

Now slowly turn your attention to the western horizon. As you turn, keep Cassiopeia in view as she immodestly anchors the rest of January's paths to the stars. Since your perspective has changed, Cass now resembles a squashed E. To her immediate left you should see a long, gentle arc of stars. Attached to the bottom of the arc is a large square turned on its corner. Many people see this as a "Great Dipper." Formally, the handle is **Andromeda**—the Chained Lady, Cassiopeia's daughter. The bottom three stars of the bowl of this "Great Dipper" form **Pegasus**—the Winged Horse. Almost everyone refers to this bowl as the **Great Square of Pegasus.**

If your sky is dark, you may be able to see the **Circlet,** the only stars I can see in **Pisces**—the Fishes.

Let's continue with our dark sky motif. The Milky Way extends from Cygnus on the northwestern horizon through Cassiopeia right to the zenith, then back to the southwestern horizon. Before you learn the constellations near the winter Milky Way, retrace your steps along the western horizon. Take another look at the Great Square of Pegasus and Andromeda. Notice again that they make up a Great Dipper.

Since it's a constellation of the Zodiac, we must pay attention to a nondescript group near Andromeda, the handle of the Great Dipper. **Aries**—the Ram is the two-star group you'll find about 20° left of Andromeda.

The Eastern Path Along the Winter Milky Way

Now face due south and lean back so you can look just west of the zenith to pick up Andromeda, the handle of the Great Dipper. It might help to lean back farther to pick up bright Cassiopeia, then drop down to Andromeda.

Perseus—the Hero is at the zenith just to the left or east of Andromeda. His head is up since you're facing south and his feet are spread in a heroic stance. (Other versions of what Perseus looks like include his holding the head of Medusa to stop a sea monster.)

Just to the left of Perseus is the bright pentagon shape of **Auriga**—the Charioteer. I have it on good authority that Charlton Heston sees a charioteer in this figure, however, most amateurs I know see the figure I've drawn, so here it is. Everyone does see the beautiful yellowish star Capella, which anchors Auriga.

Below Auriga and Perseus lie two magnificent galactic clusters, the huge **Hyades** on the left and the lovely **Pleiades** on the right. These two groups make up much of the constellation **Taurus**—the Bull. This constellation stretches all the way to the bottom star in the pentagon of Auriga. In fact, that star is within the border of Taurus, but completes too nice a pentagon even to consider drawing it as part of a fanciful bull.

Auriga and Taurus are the subject of the ZOOM map for this month. In the commentary, I'll tell you all about the exquisite galactic clusters to view in this region, as well as the Crab Nebula, a bright remnant of a supernova explosion.

Below and to the left of Taurus is the unmistakable figure of **Orion**—the Hunter. His trademark three stars mark his belt, which he wears at a rakish angle to better support his sword. The beautiful red-giant star Betelgeuse marks his right shoulder (as he sees it) and the brilliant blue-white Rigel marks his left knee.

To the left of Orion are his faithful hunting dogs **Canis Major** and **Canis Minor.** Sirius, the brightest star in the sky, is the star marking the head of Orion's Greater Dog, while Procyon is the brightest star in the Lesser Dog.

To finish up the December sky, face due east. By looking over your right shoulder, you can still see Orion as he prepares to cross the meridian. When you face east these December nights, Orion's dogs are almost on their backs. Above and to the left of Procyon, almost exactly due east, the bright stars Castor and Pollux mark the heads of **Gemini**—the Twins. Castor is the bright white star higher in this view than yellowish Pollux. These guys are also lying on their backs, but they'll be easier to see later on tonight or later in the winter at this time.

Below Gemini is another faint constellation of the Zodiac, **Cancer**—the Crab. This hard-to-see figure can almost be regarded as a placeholder between bright Gemini and **Leo**—the Lion, who's rising just above the horizon a little north of due east. If your sky is dark and your horizon low, you might be able to trace the sickle-shaped asterism that forms his head and mane. Regulus is the bright star at the bottom of the sickle.

A Recyclable Holiday Greeting

Whatever year-end holiday you celebrate, or none at all, I have this wish for you: May you use this book as the starting point on a lifelong journey of learning, enjoying, and loving the stars. May your love of the stars always bring you and yours peace and contentment. Just as the New Year brings renewal, may the stars bring you continual renewal for all of your days—and starry nights. Clear skies....

About the ZOOM Map for December

Each ZOOM map features at least one of the showpiece objects in the sky visible from the mid-northern latitudes. That's why I ZOOMed-in on the twelve regions of the sky they cover. And there are several ZOOM maps jam-packed with wonderful objects to view in binoculars or a small telescope. However, if I were compelled to pick my favorite, I'd be tempted to choose December's region: Auriga and Taurus. The reason is simple: the incomparable Pleiades.

The Lovely Seven (or 8, or 9, or 10, or is it 11?) Sisters

There are other celestial objects that are gaudier, or bigger, or brighter, or farther away, or this-er, or that-er. No object compares to the subtle beauty of the **Pleiades.** On a cold winter night, you can enjoy a view of them with just your unaided eyes, with binoculars, or with a telescope. There are no bad views of the Pleiades.

The Seven Sisters, as they were called in Greek mythology, are merely the seven brightest of about 250 stars in this bright galactic cluster. It's about 400 light-years away, and is very young, possibly only twenty million years old. Its brightest stars are hot, bluish-white powerhouses that are as much as a thousand times more luminous than our Sun. They're the stars that "live life in the fast lane," then die young.

To the unaided eye, the Pleiades are beautiful, but they're also a nice test of your eyesight. Almost everyone sees the Pleiades as a "little dipper" of seven stars. Many people see more than seven Pleiades. Some have reported more than a dozen. My usual number is ten. However many you see, I doubt you'll ever tire of seeing the Pleiades' icy blue-white stars on an inky-black background.

In binoculars, the Seven Sisters are joined by possibly seventy more stars, all icy blue-white, as the cluster is simply not old enough to have any elderly red giants. On clear, moonless, dark nights, binoculars will begin to show the nebulosity that envelopes the cluster. Studies of the spectrum of the nebulosity show that its light is reflected from the stars of the cluster. This is a reflection nebula, made up mostly of dust.

ZOOM Map for December

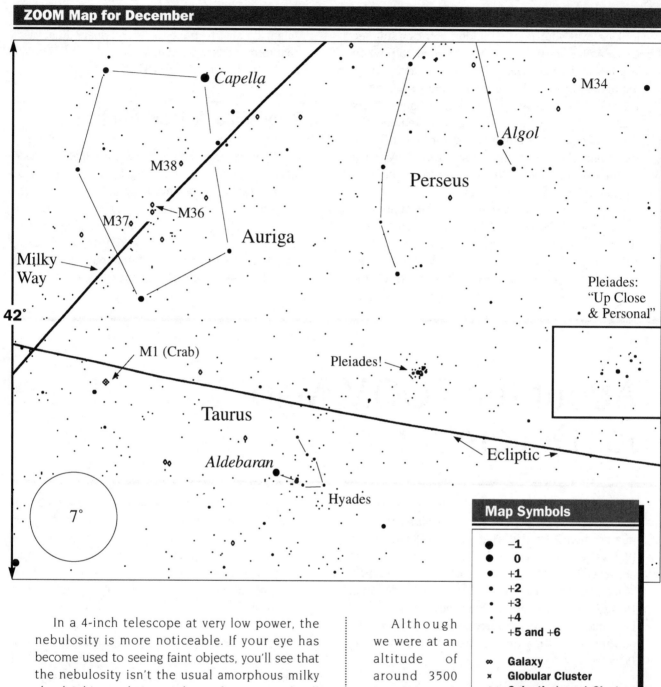

In a 4-inch telescope at very low power, the nebulosity is more noticeable. If your eye has become used to seeing faint objects, you'll see that the nebulosity isn't the usual amorphous milky cloud. It hints at being made up of many strands, all running parallel to each other. A 6-inch shows the nebula a bit better, but much larger an instrument begins to crowd things too much.

My best view of the Pleiades was in January 1986, from Shadow Mountain, California, in the high desert east of Edwards Air Force Base, where the Shuttle often lands. I had been hired by a travel agent to conduct star parties and help in seminars for people from the Los Angeles area who wanted to take a weekend to study and see Halley's Comet. We did all that, but, on that Saturday night, we also had a "night of a lifetime."

Although we were at an altitude of around 3500 feet, the night was calm and warm (47° F). The air was also bone dry. The seeing was so steady I had no trouble resolving subtle detail in galaxy M82 at 391-power in my 20-inch reflector. (I later remarked to one of my friends that "I showed little old ladies things last night that many amateur astronomers never see in their whole lifetimes!")

I was so busy coordinating the views our guests were getting through the half-dozen telescopes I didn't even look at the Pleiades until everyone was safely back on board the bus at midnight. After rolling the 20-inch into my van, and loading the rest of the heavy stuff, I stopped to look at the Pleiades, my favorites, in a 5.5-inch reflector telescope I'd only recently acquired.

I was dumbstruck at the view. (If you knew me, you would know just how remarkable such a situation would be!) At 21 power, the Pleiades were completely embedded in the wraith-like pattern of nebulosity that is so familiar in long-exposure photographs. The nebulosity seemed to show the same "iciness" that the stars themselves show. I hadn't the sense to raise the power to see the nebulosity in more detail. I watched for about an hour, only taking the occasional break to look at the Seven Sisters with only my eyes or through binoculars. I could see the nebulosity clearly without optical aid, and it was obvious in 7×50 binoculars.

At the end of the hour, the weather changed. A cold wind blew in from the northwest. The temperature dropped 10°, and the humidity shot way up. Alas, the Pleiades were, once again, "only" the Pleiades, with a "hint" of the previous nebulosity. Despite the "degradation" of my view, I stayed and watched the Seven Sisters a little longer. Even then, they were exquisite.

The Hyades (Who Shouldn't Have to Follow the Pleiades)

Just as Duke Snider was the "other" center-fielder in New York baseball in the fifties, the magnificent **Hyades** cluster is the "other" cluster in Taurus. This V-shaped group of stars is accented by the bright orangish-giant star Aldebaran, although Aldebaran is not a member of the cluster, being only half as distant.

The Hyades are only 130 light-years away, so they're one of the closest clusters visible in our sky. There are 26 stars in the Hyades brighter than magnitude 6, and 132 brighter than magnitude 9. Since the cluster is 4.5° in diameter, binoculars are the instrument of choice to see it best, as few telescopes give that wide a field.

The Hyades are possibly 400 million years old, part of a larger cluster called the Taurus Moving Group. This group is spread out over much of the constellation Taurus. It's moving toward a point in the sky near the star Betelgeuse in Orion.

The Crab Nebula (Remnants of a Supernova)

On July 4, 1054, a massive star in the constellation Taurus became a supernova. The star collapsed due to its own gravity after it ran out of nuclear fuel to sustain any type of fusion reaction in its core. The onrushing wave of material reached the center and created enormous pressure, causing a cataclysmic explosion. According to records kept by Chinese astronomers, this "guest star" was easily visible in daylight for 23 days after its initial outburst. It then diminished in brightness over a period of many months. At its peak, the supernova put out 40 percent as much energy as our entire Milky Way galaxy.

We can see the remainder of that explosion today. The **Crab Nebula, M1,** is found south of Auriga at the spot marked on the ZOOM map. Although it's usually listed as a planetary nebula, M1 is a supernova remnant, a rapidly expanding cloud of gas and dust that was blasted away from the star as it exploded. A planetary nebula, by contrast, is formed by an almost peaceful event, as the outer layer of a star is gently "sloughed off" into space.

The Crab is fairly small, 3' by 5', and that concentrates its ninth-magnitude brightness into a relatively small area. The surface brightness is fairly high. In a 4-inch at medium power, the Crab is a featureless oval, while a 6-inch begins to show the filaments of nebulosity that resemble a crab's legs, giving it its name. A 10-inch shows much more detail. The Crab Nebula is well worth standing in line to see in a telescope of the 16–24-inch class.

The Other Galactic Clusters on December's ZOOM

The whole field of December's ZOOM map is filled with galactic clusters, so it's well worth a stroll through it with binoculars on a dark winter night. Of special interest are three galactic clusters in Auriga: **M36** is a young cluster of hot bluish-white stars similar to the Pleiades, only some ten times farther away than that celebrated cluster. Also, M36 seems not to be embedded in nebulosity the way the Pleiades are. It's a fairly tightly packed group of about sixty stars. The brightest is about ninth magnitude, while the dimmest are about five magnitudes fainter.

M37 is the nicest cluster of the three in Auriga. Although it's farther than M36 and the total luminosity of its stars is not as great as M36, its stars are packed into a much smaller space, so it looks richer. M37 is a lot older than M36, so some of its stars have reached "old age" and have become red giants. The contrast of the red giants with the younger white stars is one of the primary reasons M37 is so beautiful. Adding to the beauty of M37 is that it's located in one of the richest little star clouds in the winter Milky Way.

M38 is another nice galactic cluster in Auriga. Its one hundred stars are spread over an area two-thirds the diameter of the full Moon, so it doesn't jump out at you the way M36 and M37 do. To fully appreciate M38, be sure to use a very low power. Binoculars show this cluster well..

If you propose to view M36, M37, and M38 with a small telescope, don't fail to carefully sweep the central region of Auriga's "pentagon" lying between these clusters. There are numerous other clusters there, but there are also some dark nebulas, especially between M37 and M36.

Bonus Table

Solar System Object Comparisons

Planet	How Big? (in miles)	How Massive? (Earth=1.0)	How Far? (in miles from the Sun)	"Day"	"Year"
Mercury	3,100	0.05	36.0 million	58.7 days	88.0 days
Venus	7,700	0.81	67.3 million	243.0 days	224.7 days
Earth	7,918	1.00	93.0 million	23h56m	365.3 days
Moon	2,160	0.012	239.0 thousand	27.3 days	27.3 days
Mars	4,220	0.11	141.7 million	24h37m	687.0 days
Jupiter	88,700	317.9	484.0 million	9h50m	11.9 years
Saturn	71,600	95.2	887.0 million	10h14m	129.5 years
Uranus	32,000	14.5	1,787.0 million	17h54m	84.0 years
Neptune	31,000	17.1	2,797.0 million	19h12m	164.8 years
Pluto	1,430	0.002	3,675.0 million	6.4 days	248.4 years

How Big? refers to the mean diameter of the planet. *How Far?* refers to the mean distance of the planet from the sun. The exception is the Moon. *How Far?* refers to its mean distance from Earth. *"Day"* is the time it takes for the object to rotate once on its axis, while *"Year"* refers to the time it takes to complete one revolution around the Sun. Again, figures for the Moon are relative to the Earth, not the Sun. For the record, the Sun is 864,000 miles in diameter. As massive as everything in the Solar System is, it pales in comparison to the Sun's mass. It's 740 *times* the mass of everything else!

APPENDIX

Mini-Almanac

1993

Month	Phases of the MOON 1st	Full	3rd	New	Bright PLANET Positions M	V	M	J	S	Eclipse Notes:
JANUARY	1/30	8	15	22	/	E	GEM	VIR	CAP	
FEBRUARY		6	13	21	E	E	GEM	VIR	///	
MARCH	1/31	8	15	23	M	E	GEM	VIR	CAP	**21 May** Partial SOLAR Eclipse N America, Northern Europe
APRIL	29	6	13	21	M	M	GEM	VIR	CAP	
MAY	28	6	13	21E	/	M	CNC	VIR	AQR	
JUNE	26	4E	12	20	E	M	LEO	VIR	AQR	**04 Jun** Total LUNAR Eclipse Australia, Eastern Asia
JULY	26	3	11	19	/	M	LEO	VIR	AQR	
AUGUST	24	2	10	17	M	M	VIR	VIR	AQR	
SEPTEMBER	22	1/30	9	16	E	M	VIR	VIR	CAP	**13 Nov** Partial SOLAR Eclipse N & S America
OCTOBER	22	30	8	15	E	M	LIB	///	CAP	
NOVEMBER	21	29E	7	13E	M	M	///	VIR	CAP	**29 Nov** Total LUNAR Eclipse N & S America
DECEMBER	20	28	6	13	M	/	///	VIR	CAP	

1994

Month	Phases of the MOON 3rd	New	1st	Full	Bright PLANET Positions M	V	M	J	S	Eclipse Notes
JANUARY	5	11	19	27	E	/	///	LIB	AQR	
FEBRUARY	3	10	18	26	E	/	///	LIB	///	
MARCH	4	12	20	26	M	E	AQR	LIB	AQR	
APRIL	3	11	19	25	M	E	PSC	LIB	AQR	
MAY	2	10E	18	25E	E	E	PSC	VIR	AQR	**10 May** Annular SOLAR Eclipse N America, W Europe, NW Africa
JUNE	1/30	9	16	23	E	E	ARI	VIR	AQR	
JULY	30	8	16	22	M	E	TAU	VIR	AQR	
AUGUST	29	7	14	21	/	E	GEM	LIB	AQR	**25 May** Partial LUNAR Eclipse N & S America, W Europe, Africa
SEPTEMBER	28	5	12	19	E	E	GEM	LIB	AQR	
OCTOBER	27	5	11	19	E	E	CNC	LIB	AQR	
NOVEMBER	26	3E	10	18	M	M	LEO	///	AQR	**03 Nov** Total SOLAR Eclipse S America
DECEMBER	25	2	9	18	/	M	LEO	LIB	AQR	

1995

Month	Phases of the MOON				Bright PLANET Positions					Eclipse Notes
	New	1st	Full	3rd	M	V	M	J	S	
JANUARY	1/30	8	16	24	E	M	LEO	SCO	AQR	
FEBRUARY		7	15	22	M	M	LEO	SCO	AQR	
MARCH	1/31	9	17	23	M	M	CNC	SCO	///	
APRIL	29	8	15E	22	/	M	CNC	SCO	AQR	**15 Apr** Partial LUNAR Eclipse All countries bordering Pacific Ocean except S America
MAY	29	7	14	21	E	M	LEO	SCO	AQR	
JUNE	28	6	13	19	M	M	LEO	SCO	AQR	
JULY	27	5	12	19	M	/	VIR	SCO	PSC	**29 Apr** Annular SOLAR Eclipse Central and S America
AUGUST	26	4	10	18	E	/	VIR	SCO	PSC	
SEPTEMBER	24	2	9	16	E	/	VIR	SCO	AQR	
OCTOBER	24E	1/30	8	16	M	E	LIB	SCO	AQR	**24 Oct** Total SOLAR Eclipse Asia, N Australia
NOVEMBER	22	29	7	15	/	E	SCO	SCO	AQR	
DECEMBER	22	28	7	15	E	E	SGR	///	AQR	

1996

Month	Phases of the MOON				Bright PLANET Positions					Eclipse Notes
	Full	3rd	New	1st	M	V	M	J	S	
JANUARY	5	13	20	27	E	E	SGR	SGR	AQR	
FEBRUARY	4	12	18	26	M	E	///	SGR	AQR	
MARCH	5	12	19	27	M	E	///	SGR	AQR	**04 Apr** Total LUNAR Eclipse W Asia, Africa, Europe, E N America, S America
APRIL	4E	10	17E	25	E	E	///	SGR	///	
MAY	3	10	17	25	/	E	ARI	SGR	PSC	
JUNE	1	8	16	24	M	/	TAU	SGR	PSC	**10 Apr** Partial SOLAR Eclipse S Pacific Ocean
JULY	1/30	7	15	23	/	M	TAU	SGR	PSC	
AUGUST	28	6	14	22	E	M	GEM	SGR	PSC	**27 Sep** Total LUNAR Eclipse N & S America, Africa, Europe
SEPTEMBER	27E	4	12	20	E	M	CNC	SGR	PSC	
OCTOBER	26	4	12E	19	M	M	LEO	SGR	PSC	
NOVEMBER	25	3	11	18	/	M	LEO	SGR	PSC	**12 Oct** Partial SOLAR Eclipse NE N America, Europe, N Africa
DECEMBER	24	3	10	17	E	M	LEO	SGR	PSC	

1997

Month	Phases of the MOON				Bright PLANET Positions					Eclipse Notes
	3rd	New	1st	Full	M	V	M	J	S	
JANUARY	2/31	9	15	23	M	/	VIR	///	PSC	
FEBRUARY		7	14	22	M	M	VIR	SGR	PSC	
MARCH	2/31	9E	16	24E	E	/	LEO	CAP	PSC	**09 Mar** Total SOLAR Eclipse NE Asia, NW Pacific Ocean
APRIL	30	7	14	22	E	/	LEO	CAP	PSC	
MAY	29	6	14	22	M	E	LEO	CAP	PSC	**24 Mar** Partial LUNAR Eclipse W Africa, W Europe, N & S America
JUNE	27	5	13	22	M	E	VIR	CAP	PSC	
JULY	26	4	12	20	E	E	VIR	CAP	PSC	
AUGUST	25	3	11	18	E	E	VIR	CAP	PSC	**02 Sep** Partial SOLAR Eclipse Australia, SW Pacific Ocean
SEPTEMBER	23	2E	10	16E	M	E	LIB	CAP	PSC	
OCTOBER	23	1/31	9	16	/	E	SCO	CAP	PSC	
NOVEMBER	23	30	7	14	E	E	SGR	CAP	PSC	**16 Sep** Total LUNAR Eclipse Asia, Europe, Africa, Australia
DECEMBER	21	29	7	14	E	E	SGR	CAP	PSC	

1998

Month	Phases of the MOON				Bright PLANET Positions					Eclipse Notes
	1st	Full	3rd	New	M	V	M	J	S	
JANUARY	5	12	20	28	M	/	CAP	CAP	PSC	
FEBRUARY	3	11	19	26E	/	M	AQR	///	PSC	**26 Feb** Total SOLAR Eclipse S N America, N S America
MARCH	5	13	21	28	E	M	PSC	AQR	PSC	
APRIL	3	11	19	26	M	M	///	AQR	///	
MAY	3	11	19	25	M	M	///	AQR	///	
JUNE	2	10	17	24	E	M	///	PSC	ARI	
JULY	1/31	9	16	23	E	M	GEM	PSC	ARI	
AUGUST	30	8	14	22E	M	M	GEM/CNC	PSC	ARI	**22 Aug** Annular SOLAR Eclipse SE Asia, Australia, SW Pacific Ocean
SEPTEMBER	28	6	13	20	M	M	LEO	PSC	ARI	
OCTOBER	28	5	12	20	E	/	LEO	AQR	ARI	
NOVEMBER	27	4	11	19	E	/	LEO	AQR	ARI	
DECEMBER	26	3	10	18	M	E	VIR	AQR	ARI	

1999

Month	Phases of the MOON				Bright PLANET Positions					Eclipse Notes
	Full	3rd	New	1st	M	V	M	J	S	
JANUARY	2/31	9	17	24	M	E	VIR	PSC	AQR	
FEBRUARY		8	16E	23	E	E	VIR	PSC	PSC	
MARCH	2/31	10	17	24	E	E	LIB	PSC	ARI	
APRIL	30	9	16	22	M	E	LIB	PSC	ARI	**16 Feb** annular SOLAR Eclipse S Africa, Australia, SE Asia
MAY	30	8	15	22	M	E	VIR	PSC	ARI	
JUNE	28	7	13	20	E	E	VIR	PSC	ARI	**28 Jul** Partial LUNAR Eclipse Australia, E Asia, NW N America
JULY	28E	6	13	20	E	E	VIR	ARI	ARI	
AUGUST	26	4	11E	19	M	E	LIB	ARI	ARI	**11 Aug** Total SOLAR Eclipse E N America, Europe, N Africa, W Asia
SEPTEMBER	25	2	9	17	E	M	SCO	ARI	ARI	
OCTOBER	24	2/31	9	17	E	M	SGR	ARI	ARI	
NOVEMBER	23	29	8	16	E	M	SGR	ARI	ARI	
DECEMBER	22	29	7	16	M	M	CAP	PAC	ARI	

2000

Month	Phases of the MOON				Bright PLANET Positions					Eclipse Notes
	New	1st	Full	3d	M	V	M	J	S	
JANUARY	6	14	21E	28	/	M	AQR	PSC	ARI	**21 Jan** Total LUNAR Eclipse N & S America, W Africa, Europe
FEBRUARY	5E	12	19	27	E	M	PSC	ARI	ARI	
MARCH	6	13	20	28	M	M	PSC	ARI	ARI	**05 Feb** Partial SOLAR Eclipse Antarctica, S Polar Regions
APRIL	4	11	18	26	M	M	ARI	ARI	ARI	
MAY	4	10	18	26	E	/	TAU	///	///	**16 Jul** Total LUNAR Eclipse Australia, E Asia
JUNE	2	9	16	25	E	/	///	TAU	TAU	
JULY	1/31E	8	16E	24	M	E	///	TAU	TAU	**31 Jul** Partial SOLAR Eclipse N Asia, NW N America
AUGUST	29	7	15	22	/	E	CNC	TAU	TAU	
SEPTEMBER	27	5	13	21	E	E	LEO	TAU	TAU	**25 Dec** Partial SOLAR Eclipse N & Central America
OCTOBER	27	5	13	20	E	E	LEO	TAU	TAU	
NOVEMBER	25	4	11	18	M	E	VIR	TAU	TAU	
DECEMBER	25E	4	11	18	/	E	VIR	TAU	TAU	

SUNRISE-SUNSET TABLE for Latitudes 30°–50° North

LATITUDE: DATE	30° N rise–set	35° N rise–set	40° N rise–set	45° N rise–set	50° N rise–set
JAN 1	0655–1711	0707–1659	0721–1645	0738–1621	0758–1608
11	0657–1719	0708–1707	0722–1654	0737–1638	0756–1620
21	0655–1727	0706–1717	0718–1705	0730–1652	0748–1635
FEB 1	0651–1736	0700–1728	0709–1718	0721–1705	0735–1635
11	0644–1745	0651–1738	0659–1730	0708–1719	0719–1710
21	0635–1753	0640–1748	0645–1742	0654–1735	0700–1727
MAR 1	0626–1800	0629–1756	0633–1752	0637–1747	0643–1643
11	0614–1806	0616–1804	0618–1803	0619–1801	0622–1759
21	0602–1812	0602–1812	0601–1813	0601–1814	0600–1815
APR 1	0051–1819	0549–1822	0546–1824	0541–1828	0537–1831
11	0537–1824	0533–1829	0528–1835	0523–1840	0515–1848
21	0527–1831	0516–1841	0508–1845	0504–1854	0455–1904
MAY 1	0517–1838	0509–1845	0500–1855	0449–1904	0436–1919
11	0509–1844	0456–1854	0449–1905	0436–1918	0419–1934
21	0502–1850	0451–1901	0438–1914	0425–1930	0406–1948
JUN 1	0500–1856	0448–1909	0434–1922	0417–1939	0356–1959
11	0458–1901	0441–1914	0431–1929	0413–1947	0351–2009
21	0500–1904	0417–1917	0432–1932	0414–1950	0351–2012
JUL 1	0502–1905	0450–1918	0435–1933	0417–1951	0355–2012
11	0507–1903	0455–1915	0441–1929	0425–1947	0403–2006
21	0512–1900	0501–1911	0449–1924	0434–1938	0415–1957
AUG 1	0519–1853	0509–1903	0458–1913	0446–1926	0429–1942
11	0525–1845	0517–1852	0508–1901	0457–1916	0444–1925
21	0531–1835	0525–1841	0517–1848	0509–1856	0459–1905
SEP 1	0537–1822	0533–1826	0529–1831	0521–1835	0515–1844
11	0542–1810	0540–1812	0537–1815	0534–1818	0530–1822
21	0548–1757	0547–1758	0547–1759	0546–1800	0545–1800
OCT 1	0553–1745	0555–1744	0556–1742	0559–1740	0600–1738
11	0600–1733	0603–1729	0606–1725	0611–1721	0616–1717
21	0606–1723	0611–1718	0617–1712	0624–1702	0632–1657
NOV 1	0614–1712	0621–1706	0629–1657	0639–1649	0651–1636
11	0622–1706	0631–1657	0641–1647	0654–1635	0707–1620
21	0631–1701	0641–1651	0652–1639	0706–1623	0723–1609
DEC 1	0639–1700	0650–1649	0703–1635	0718–1619	0737–1601
11	0646–1700	0658–1648	0712–1635	0729–1618	0749–1558
21	0651–1704	0704–1652	0718–1638	0735–1621	0756-1600

HOW TO USE THE SUNRISE-SUNSET TABLE:

As I've done with all the times in this book, these times are STANDARD TIME for your central meridian. If you live on or near your central meridian, these times will be accurate to within a minute or two. Subtract 4-minutes for each degree you live East of your standard meridian. Add 4-minutes for each degree you live west of your standard meridian. Add 1-hour to times on this table for DAYLIGHT SAVINGS TIME.

This is not a perfectly precise table. It's only designed to help you estimate times within a few minutes, so don't quote it in a lawsuit! To estimate the times that lie between the table entries, "split the difference". If the date or latitude isn't exactly between two entries, offset your estimate a little.

What Map to Use When

DATE	TIME													
	1720	1820	1920	2020	2120	2220	2320	0020	0120	0220	0320	0420	0520	0620
JAN 1	OCT		NOV		DEC		JAN		FEB		MAR		APR	
15		NOV		DEC		JAN		FEB		MAR		APR		MAY
FEB 1	NOV		DEC		JAN		FEB		MAR		APR		MAY	
15		DEC		JAN		FEB		MAR		APR		MAY		JUN
MAR 1	DEC		JAN		FEB		MAR		APR		MAY		JUN	
15		JAN		FEB		MAR		APR		MAY		JUN		JUL
APR 1	JAN		FEB		MAR		APR		MAY		JUN		JUL	
15		FEB		MAR		APR		MAY		JUN		JUL		AUG
MAY 1	FEB		MAR		APR		MAY		JUN		JUL		AUG	
15		MAR		APR		MAY		JUN		JUL		AUG		SEP
JUN 1	MAR		APR		MAY		JUN		JUL		AUG		SEP	
15		APR		MAY		JUN		JUL		AUG		SEP		OCT
JUL 1	APR		MAY		JUN		JUL		AUG		SEP		OCT	
15		MAY		JUN		JUL		AUG		SEP		OCT		NOV
AUG 1	MAY		JUN		JUL		AUG		SEP		OCT		NOV	
15		JUN		JUL		AUG		SEP		OCT		NOV		DEC
SEP 1	JUN		JUL		AUG		SEP		OCT		NOV		DEC	
15		JUL		AUG		SEP		OCT		NOV		DEC		JAN
OCT 1	JUL		AUG		SEP		OCT		NOV		DEC		JAN	
15		AUG		SEP		OCT		NOV		DEC		JAN		FEB
NOV 1	AUG		SEP		OCT		NOV		DEC		JAN		FEB	
15		SEP		OCT		NOV		DEC		JAN		FEB		MAR
DEC 1	SEP		OCT		NOV		DEC		JAN		FEB		MAR	
15		OCT		NOV		DEC		JAN		FEB		MAR		APR

HOW TO READ THIS TABLE:

To make it easier to format this table, all times are written in the 24-hour style that the military, transportation, and space agencies use. Using this type of notation also makes it easier to do time calculations. So, 1720 is 5:20 P.M., and 0020 is 12:20 A.M. Also, as with all other times in this book, the times you find in this table are STANDARD TIME. Add 1-hour to obtain DAYLIGHT SAVINGS TIME.

Index